Antenna Toolkit

Antenna Toolkit 2nd Edition

Joseph J. Carr, K4IPV

OXFORD AMSTERDAM BOSTON LONDON NEW YORK PARIS
SAN DIEGO SAN FRANCISCO SINGAPORE SYDNEY TOKYO

Newnes
An imprint of Elsevier Science
Linacre House, Jordan Hill, Oxford OX2 8DP
200 Wheeler Road, Burlington, MA 01803

First published 1997
Reprinted 1998
Second edition 2001
Reprinted 2003

Copyright © 1997, 2001, Joseph J. Carr. All rights reserved

No part of this publication may be reproduced in any material form
(including photocopying or storing in any medium by electronic means
and whether or not transiently or incidentally to some other use of this
publication) without the written permission of the copyright holder except
in accordance with the provisions of the Copyright, Designs and Patents
Act 1988 or under the terms of a licence issued by the Copyright Licensing
Agency Ltd, 90 Tottenham Court Road, London, England W1T 4LP.
Applications for the copyright holder's written permission to reproduce
any part of this publication should be addressed to the publisher.
Permissions may be sought directly from Elsevier's Science and Technology
Rights Department in Oxford, UK; phone: (+44) (0) 1865 843830;
fax: (+44) (0) 1865 853333; e-mail: permissions@elsevier.co.uk. You may also
complete your request on-line via the Elsevier Science homepage
(http://www.elsevier.com), by selecting 'Customer Support' and then
'Obtaining Permissions'

British Library Cataloguing in Publication Data
A catalogue record for this book is available from the British Library

Library of Congress Cataloguing in Publication Data
A catalogue record for this book is available from the Library of Congress

ISBN 0 7506 4947 X

For information on all Newnes publications
visit our website at www.newnespress.com

Typeset by Keyword
Printed and bound in Great Britain by Biddles Ltd, *www.biddles.co.uk*

Contents

Preface   vii

1. Radio signals on the move   1
2. Antenna basics   19
3. Wire, connection, grounds, and all that   49
4. Marconi and other unbalanced antennas   69
5. Doublets, dipoles, and other Hertzian antennas   87
6. Limited space antennas   118
7. Large loop antennas   129
8. Wire array antennas   153
9. Small loop antennas   176
10. Yagi beam antennas   195
11. Impedance matching   203
12. Simple antenna instrumentation and measurements   221
13. Getting a 'good ground'   237

Index   249

Preface

If you are interested in amateur radio, short-wave listening, scanner monitoring, or any other radio hobby, then you will probably need to know a few things about radio antennas. This book is intended for the radio enthusiast – whether ham operator, listening hobbyist, or radio science observer – who wants to build and use antennas for their particular requirements and location. All of the antennas in this book can be made from wire, even though it is possible to use other materials if you desire.

These antennas have several advantages. One of the most attractive is that they can provide decent performance on the cheap. As one who has lived through the experience of being broke, I learned early to use bits of scrap wire to get on the air. My first novice antenna back in the late 1950s was a real patched-together job – but it worked really well (or so I thought at the time!).

Another advantage of wire antennas is that they are usually quite easy to install. A couple of elevated supports (tree, roof, mast), a few meters of wire, a few bits of radio hardware, and you are in the business of putting up an antenna. As long as you select a safe location, then you should have little difficulty erecting that antenna.

Finally, most high-frequency (HF) short-wave antennas are really easy to get working properly. One does not need to be a rocket scientist – or professional antenna rigger – to make most of these antennas perform as well as possible with only a little effort. There is quite a bit of detailed technical material to digest if you want to be a professional antenna engineer, but you can have good results if you follow a few simple guidelines.

SOFTWARE SUPPLEMENT TO THIS BOOK

At the time this book was conceived it was noted that the technology now exists to make Microsoft Windows-based antenna software available to readers along with the book. The software can be used to calculate the dimensions of the elements of most of the antennas in this book, as well as a few that are not. There are also some graphics in the software that show you a little bit about antenna hardware, antenna construction, and the like.

ANTENNA SAFETY

Every time I write about antenna construction I talk a little bit about safety. The issue never seems too old or too stale. The reason is that there seem to be plenty of people out there who never get the word. Antenna erection does not have to be dangerous, but if you do it wrong it can be very hazardous. Antennas are deceptive because they are usually quite lightweight, and can easily be lifted. I have no trouble lifting my trap vertical and holding it aloft – on a windless day. But if even a little wind is blowing (and it almost always is), then the 'sail area' of the antenna makes it a lot 'heavier' (or so it seems). Always use a buddy-system when erecting antennas. I have a bad back caused by not following my own advice.

Another issue is electrical safety. Do not ever, ever, ever toss an antenna wire over the power lines. Ever. Period. Also, whatever type of antenna you put up, make sure that it is in a location where it cannot possibly fall over and hit the power line.

The last issue is to be careful when digging to lay down radials. You really do not want to hit water lines, sewer lines, buried electrical service lines, or gas lines. I even know of one property where a long-distance oil pipeline runs beneath the surface. If you do not know where these lines are, try to guess by looking at the locations of the meters on the street, and the service entrance at the house. Hint: most surveyors' plans (those map-like papers you get at settlement) show the location of the buried services. They should also be on maps held by the local government (although you might have to go to two or three offices!. The utility companies can also help.

A NOTE ABOUT UNITS AND PRACTICES

This book was written for an international readership, even though I am American. As a result, some of the material is written in terms of US standard practice. Wherever possible, I have included UK standard wire sizes and metric units. Metric units are not in common usage in the USA, but rather we still use the old English system of feet, yards, and inches. Although many Americans (including myself) wish the USA would convert to SI units, it is not likely in the near future. UK readers with a sense of history might recognize why this might be true – as you may recall from the George III unpleasantness, Americans do not like foreign rulers, so it is not likely that our measuring rulers will be marked in centimeters rather than inches.* For those who have not yet mastered the intricacies of converting between the two systems:

> 1 inch = 2.54 centimeters (cm) = 25.4 millimeters (mm)
> 1 foot = 30.48 cm = 0.3048 meter (m)
> 1 m = 39.37 inches = 3.28 feet

Joseph J. Carr

---

*I apologize for the bad play on words, but I could not help it.

CHAPTER 1

Radio signals on the move

Anyone who does any listening to radio receivers at all – whether as a ham operator, a short-wave listener, or scanner enthusiast – notices rather quickly that radio signal propagation varies with time and something mysterious usually called 'conditions.' The rules of radio signal propagation are well known (the general outlines were understood in the late 1920s), and some predictions can be made (at least in general terms). Listen to almost any band, and propagation changes can be seen. Today, one can find propagation predictions in radio magazines, or make them yourself using any of several computer programs offered in radio magazine advertisements. Two very popular programs are any of several versions of IONCAP, and a Microsoft Windows program written by the Voice of America engineering staff called VOACAP.

Some odd things occur on the air. For example, one of my favorite local AM broadcast stations broadcasts on 630 kHz. During the day, I get interference-free reception. But after the Sun goes down, the situation changes radically. Even though the station transmits the same power level, it fades into the background din as stations to the west and south of us start skipping into my area. The desired station still operates at the same power level, but is barely audible even though it is only 20 miles (30 km) away.

Another easily seen example is the 3–30 MHz short-wave bands. Indeed, even those bands behave very differently from one another. The lower-frequency bands are basically ground wave bands during the day, and become long-distance 'sky wave' bands at night (similar to the AM broadcast band (BCB)). Higher short-wave bands act just the opposite: during the day they are long-distance 'skip' bands, but some time after sunset, become ground wave bands only.

The very high-frequency/ultra high-frequency (VHF/UHF) scanner bands are somewhat more consistent than the lower-frequency bands. But even in those bands sporadic-E skip, meteor scatter, and a number of other phenomena cause propagation anomalies. In the scanner bands there are summer and winter differences in heavily vegetated regions that are attributed to the absorptive properties of the foliage. I believe I experienced that phenomenon using my 2 m ham radio rig in the simplex mode (repeater operation can obscure the effect due to antenna and location height).

THE EARTH'S ATMOSPHERE

Electromagnetic waves do not need an atmosphere in order to propagate, as you will undoubtedly realize from the fact that space vehicles can transmit radio signals back to Earth in a near vacuum. But when a radio wave does propagate in the Earth's atmosphere, it interacts with the atmosphere, and its path of propagation is altered. A number of factors affect the interaction, but it is possible to break the atmosphere into several different regions according to their respective effects on radio signals.

The atmosphere, which consists largely of oxygen (O_2) and nitrogen (N_2) gases, is broken into three major zones: the **troposphere**, **stratosphere**, and **ionosphere** (Figure 1.1). The boundaries between these regions are not very well defined, and change both diurnally (i.e. over the course of a day) and seasonally.

The troposphere occupies the space between the Earth's surface and an altitude of 6–11 km. The temperature of the air in the troposphere varies with altitude, becoming considerably lower at high altitude compared with ground temperature. For example, a +10°C surface temperature could reduce to −55°C at the upper edges of the troposphere.

The stratosphere begins at the upper boundary of the troposphere (6–11 km), and extends up to the ionosphere (≈50 km). The stratosphere is called an **isothermal region** because the temperature in this region is relatively constant despite altitude changes.

The ionosphere begins at an altitude of about 50 km and extends up to 500 km or so. The ionosphere is a region of very thin atmosphere. Cosmic rays, electromagnetic radiation of various types (including ultraviolet light from the Sun), and atomic particle radiation from space (most of it from the Sun), has sufficient energy to strip electrons away from the gas molecules of the atmosphere. The O_2 and N_2 molecules that lost electrons are called **positive ions**. Because the density of the air is so low at those altitudes, the ions and electrons can travel long distances before neutralizing each other

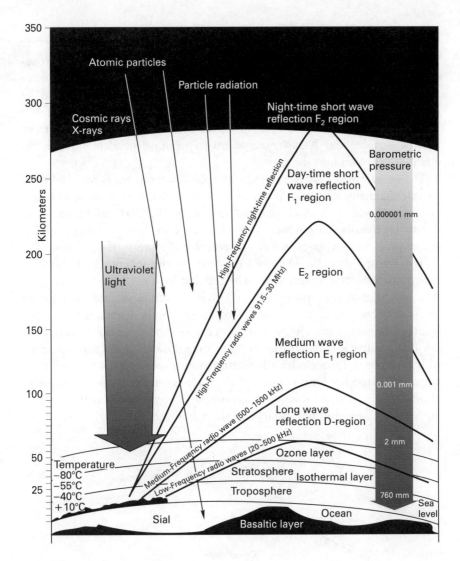

FIGURE 1.1

by recombining. Radio propagation on some bands varies markedly between daytime and night-time because the Sun keeps the level of ionization high during daylight hours, but the ionization begins to fall off rapidly after sunset, altering the radio propagation characteristics after dark. The ionization does not occur at lower altitudes because the air density is such that the positive ions and free electrons are numerous and close together, so recombination occurs rapidly.

PROPAGATION PATHS

There are four major propagation paths: **surface wave**, **space wave**, **tropospheric**, and **ionospheric**. The ionospheric path is important to medium-wave and HF propagation, but is not important to VHF, UHF, or microwave propagation. The space wave and surface wave are both **ground waves**, but behave differently. The surface wave travels in direct contact with the Earth's surface, and it suffers a severe frequency-dependent attenuation due to absorption into the ground.

The space wave is also a ground wave phenomenon, but is radiated from an antenna many wavelengths above the surface. No part of the space wave normally travels in contact with the surface; VHF, UHF, and microwave signals are usually space waves. There are, however, two components of the space wave in many cases: **direct** and **reflected** (Figure 1.2).

The ionosphere is the region of the Earth's atmosphere that is between the stratosphere and outer space. The peculiar feature of the ionosphere is that molecules of atmospheric gases (O_2 and N_2) can be ionized by stripping away electrons under the influence of solar radiation and certain other sources of energy (see Figure 1.1). In the ionosphere the air density is so low that positive ions can travel relatively long distances before recombining with electrons to form electrically neutral atoms. As a result, the ionosphere remains ionized for long periods of the day – even after sunset. At lower altitudes, however, air density is greater, and recombination thus occurs rapidly. At those altitudes, solar ionization diminishes to nearly zero imme-

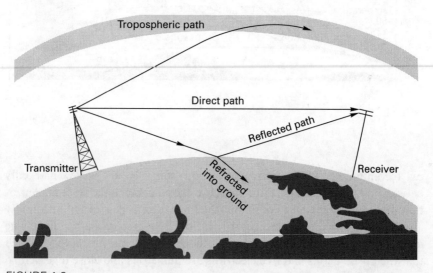

FIGURE 1.2

diately after sunset or never achieves any significant levels even at local noon.

Ionization and recombination phenomena in the ionosphere add to the noise level experienced at VHF, UHF, and microwave frequencies. The properties of the ionosphere are therefore important at these frequencies because of the noise contribution. In addition, in satellite communications there are some transionospheric effects.

GROUND WAVE PROPAGATION

The ground wave, naturally enough, travels along the ground, or at least in close proximity to it (Figure 1.3).

There are two basic forms of ground wave: **space wave** and **surface wave**. The space wave does not actually touch the ground. As a result, space wave attenuation with distance in clear weather is about the same as in free space (except above about 10 GHz, where absorption by H_2O and O_2 increases dramatically). Of course, above the VHF region, weather conditions add attenuation not found in outer space.

The surface wave is subject to the same attenuation factors as the space wave, but in addition it also suffers ground losses. These losses are due to ohmic resistive losses in the conductive earth. Surface wave attenuation is a function of frequency, and increases rapidly as frequency increases. For both of these forms of ground wave, communications is affected by the following factors: *wavelength*, *height of both receive and transmit antennas*, *distance between antennas*, and *terrain and weather along the transmission path*.

Ground wave communications also suffer another difficulty, especially at VHF, UHF, and microwave frequencies. The space wave is like a surface wave, but is radiated many wavelengths above the surface. It is made up of

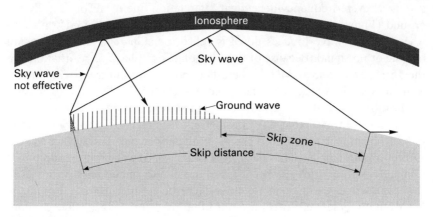

FIGURE 1.3

two components (see Figure 1.2): **direct** and **reflected** waves. If both of these components arrive at the receive antenna they will add algebraically to either increase or decrease signal strength. There is nearly always a phase shift between the two components because the two signal paths have different lengths. In addition, there may be a 180° (π radians) phase reversal at the point of reflection (especially if the incident signal is horizontally polarized).

Multipath phenomena exist because of interference between the direct and reflected components of the space wave. The form of multipath phenomenon that is, perhaps, most familiar to many readers (at least those old enough to be 'pre-cable') is ghosting in television reception. Some multipath events are transitory in nature (as when an aircraft flies through the transmission path), while others are permanent (as when a large building or hill reflects the signal). In mobile communications, multipath phenomena are responsible for reception dead zones and 'picket fencing.' A dead zone exists when destructive interference between direct and reflected (or multiple reflected) waves drastically reduces signal strengths. This problem is most often noticed at VHF and above when the vehicle is stopped; and the solution is to move the antenna a quarter wavelength. Picket fencing occurs as a mobile unit moves through successive dead zones and signal enhancement (or normal) zones, and sounds like a series of short noise bursts.

At VHF, UHF, and microwave frequencies the space wave is limited to so-called 'line of sight' distances. The horizon is theoretically the limit of communications distance, but the radio horizon is actually about 15% further than the optical horizon. This phenomenon is due to refractive bending in the atmosphere around the curvature of the Earth, and makes the geometry of the situation look as if the Earth's radius is four-thirds the actual radius.

The surface wave travels in direct contact with the Earth's surface, and it suffers a severe frequency-dependent attenuation due to absorption by the ground (Figure 1.3). The zone between the end of the ground wave and where the sky wave touches down is called the **skip zone**, and is a region of little or no signal. Because of this phenomenon, I have seen situations on the 15m band (21.390 MHz) where two stations 65km apart (Baltimore, Maryland, and Fairfax, Virginia) could not hear each other, and their communications have to be relayed via a ham station in Lima, Peru!

The surface wave extends to considerable heights above the ground level, although its intensity drops off rapidly at the upper end. The surface wave is subject to the same attenuation factors as the space wave, but in addition it also suffers ground losses. These losses are due to ohmic resistive losses in the conductive earth, and to the dielectric properties of the Earth. Horizontally polarized waves are not often used for surface wave communications because the Earth tends to short circuit the electrical (E) field component. On vertically polarized waves, however, the Earth offers electrical resistance to the *E*-field and returns currents to following waves. The conductivity of the soil determines how much energy is returned.

IONOSPHERIC PROPAGATION

Now let us turn our attention to the phenomena of **skip communications** as seen in the short-wave bands, plus portions of the medium-wave and lower VHF regions. Ionospheric propagation is responsible for intercontinental broadcasting and communications.

Long-distance radio transmission is carried out on the HF bands (3–30 MHz), also called the 'short-wave' bands. These frequencies are used because of the phenomenon called **skip**. Under this type of propagation the Earth's ionosphere acts as if it is a 'radio mirror,' to reflect the signal back to Earth. This signal is called the **sky wave**. Although the actual phenomenon is based on refraction (not reflection, as is frequently believed) the appearance to the casual ground observer is that short-wave and low-VHF radio signals are reflected from the ionosphere as if it were a kind of radio mirror. The actual situation is a little different, but we will deal with that issue in a moment.

The key lies in the fact that a seeming radio mirror is produced by ionization of the upper atmosphere. The upper portion of the atmosphere is called the 'ionosphere' because it tends to be easily **ionized** by solar and cosmic radiation phenomena. The reason for the ease with which that region (50–500 km above the surface) ionizes is that the air density is very low. Energy from the Sun strips away electrons from the outer shells of oxygen and nitrogen molecules, forming free electrons and positive ions. Because the air is so rarified at those altitudes, these charged particles can travel great distances before recombining to form electrically neutral atoms again. As a result, the average ionization level remains high in that region.

Several sources of energy will cause ionization of the upper atmosphere. Cosmic radiation from outer space causes some degree of ionization, but the majority of ionization is caused by solar energy. The role of cosmic radiation was first noticed during World War II when military radar operators discovered that the distance at which their equipment could detect enemy aircraft was dependent upon whether or not the Milky Way was above the horizon (although it was theorized 10 years earlier). Intergalactic radiation raised the background microwave noise level, thereby adversely affecting the signal-to-noise ratio.

The ionosphere is divided for purposes of radio propagation studies into various layers that have different properties. These layers are only well defined in textbooks, however, and even there we find a variation in the height above the Earth's surface where these layers are found. In addition, the real physical situation is such that layers do not have sharply defined boundaries, but rather fade one into another. The division into layers is therefore somewhat arbitrary. These layers (shown earlier in Figure 1.1) are designated D, E, and F (with F being further subdivided into the F_1 and F_2 sublayers).

D-layer

The D-layer is the lowest layer in the ionosphere, and exists from approximately 50 to 90 km above the Earth's surface. This layer is not ionized as much as higher layers because all forms of solar energy that cause ionization are severely attenuated by the higher layers above the D-layer. Another reason is that the D-layer is much denser than the E- and F-layers, and that density of air molecules allows ions and electrons to recombine to form electroneutral atoms very quickly.

The extent of D-layer ionization is roughly proportional to the height of the Sun above the horizon, so will achieve maximum intensity at midday. The D-layer exists mostly during the warmer months of the year because of both greater height of the sun above the horizon and the longer hours of daylight. The D-layer almost completely disappears after local sunset, although some observers have reported sporadic incidents of D-layer activity for a considerable time past sunset. The D-layer exhibits a large amount of absorption of medium-wave and short-wave signals, to such an extent that signals below 4–6 MHz are completely absorbed by the D-layer.

E-layer

The E-layer exists at altitudes between approximately 100 and 125 km. Instead of acting as an attenuator it acts primarily as a reflector although signals do undergo a degree of attenuation.

Like the D-layer, ionization in this region only exists during daylight hours, peaking around midday and falling rapidly after sunset. After nightfall the layer virtually disappears although there is some residual ionization there during the night-time hours.

The distance that is generally accepted to be maximum that can be achieved using E-layer propogation is 2500 km, although it is generally much less than this and can be as little as 200 km.

One interesting and exciting aspect of this region is a phenomenon called Es or sporadic E. When this occurs a layer or cloud of very intense ionization forms. This can reflect signals well into the VHH region of the radio spectrum. Although generally short lived, there can be openings on bands as high as 2 meters (144 MHz). These may last as little as a few minutes, whilst long openings may last up to a couple of hours. The phenomenon also affects lower frequencies like the 10 meter and 6 meter amateur bands as well as the VHF FM band. Sporadic E is most common in the summer months, peaking in June (in the northern hemisphere). Distances of between 1000 and 2500 km can be reached using this mode of propagation.

F-layer

The F-layer of the ionosphere is the region that is the principal cause of long-distance short-wave communications. This layer is located from about 150–500 km above the Earth's surface. Unlike the lower layers, the air density in the F-layer is low enough that ionization levels remain high all day, and decay slowly after local sunset. Minimum levels are reached just prior to local sunrise. Propagation in the F-layer is capable of skip distances up to 4000 km in a single hop. During the day there are actually two identifiable, distinct sublayers in the F-layer region, and these are designated the 'F_1' and 'F_2' layers. The F_1 layer is found approximately 150–250 km above the Earth's surface, while the F_2 layer is above the F_1 to the 450–500 km limit. Beginning at local sundown, however, the lower regions of the F_1 layer begin to de-ionize due to recombination of positive ions and free electrons. At some time after local sunset the F_1 and F_2 layers have effectively merged to become a single reduced layer beginning at about 300 km.

The height and degree of ionization of the F_2 layer varies over the course of the day, with the season of the year, and with the 27 day cycle of the sun. The F_2 layer begins to form shortly after local sunrise, and reaches a maximum shortly before noon. During the afternoon the F_2 layer ionization begins to decay in an exponential manner until, for purposes of radio propagation, it disappears sometime after local sunset. There is some evidence that ionization in the F-layer does not completely disappear, but its importance to HF radio communication does disappear.

IONOSPHERIC VARIATION AND DISTURBANCES

The ionosphere is an extremely dynamic region of the atmosphere, especially from a radio operator's point of view, for it significantly alters radio propagation. The dynamics of the ionosphere are conveniently divided into two general classes: **regular variation** and **disturbances**. We will now look at both types of ionospheric change.

Ionospheric variation

There are several different forms of variation seen on a regular basis in the ionosphere: **diurnal**, **27 day** (monthly), **seasonal**, and **11 year cycle**.

Diurnal (daily) variation

The Sun rises and falls in a 24 hour cycle, and because it is a principal source of ionization of the upper atmosphere, one can expect diurnal variation. During daylight hours the E- and D-levels exist, but these disappear at night. The height of the F_2 layer increases until midday, and then decreases until evening, when it disappears or merges with other layers. As a result of higher absorption in the E- and D-layers, lower frequencies are not useful during daylight hours. On the other hand, the F-layers reflect higher frequencies during the day. In the 1–30 MHz region, higher frequencies (>11 MHz) are used during daylight hours and lower frequencies (<11 MHz) at night. Figure 1.4B shows the number of sunspots per year since 1700.

27 day cycle

Approximately monthly in duration, this variation is due to the rotational period of the Sun. Sunspots (Figure 1.4A) are localized on the surface of the Sun, so will face the Earth only during a portion of the month. As new sunspots are formed, they do not show up on the earthside face until their region of the Sun rotates earthside.

Seasonal cycle

The Earth's tilt varies the exposure of the planet to the Sun on a seasonal basis. In addition, the Earth's yearly orbit is not circular, but elliptical. As a result, the intensity of the Sun's energy that ionizes the upper atmosphere varies with the seasons of the year. In general, the E-, D-, and F-layers are affected, although the F_2 layer is only minimally affected. Ion density in the F_2 layer tends to be highest in winter, and less in summer. During the summer, the distinction between F_1 and F_2 layers is less obvious.

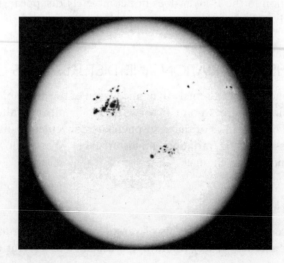

FIGURE 1.4A

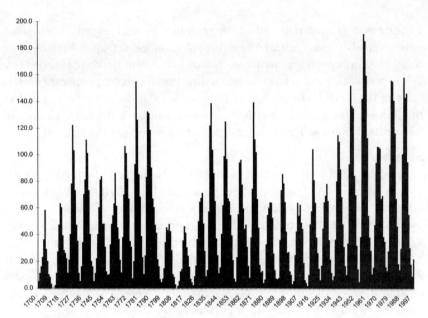

FIGURE 1.4B

11 year cycle

The number of sunspots, statistically averaged, varies on an approximately 11 year cycle (Fig. 1.4B). As a result, the ionospheric effects that affect radio propagation also vary on an 11 year cycle. Radio propagation in the short-wave bands is best when the average number of sunspots is highest. Peaks occurred in 1957, 1968, 1979, and 1990.

Events on the surface of the Sun sometimes cause the radio mirror to seem almost perfect, and make spectacular propagation possible. At other times, however, solar disturbances disrupt radio communications for days at a time.

There are two principal forms of solar energy that affect short-wave communications: **electromagnetic radiation** and **charged solar particles**. Most of the radiation is beyond the visible spectrum, in the ultraviolet and X-ray/γ-ray region of the spectrum. Because electromagnetic radiation travels at the speed of light, solar events that release radiation cause changes to the ionosphere about 8 minutes later. Charged particles, on the other hand, have a finite mass and so travel at a considerably slower velocity. They require 2 or 3 days to reach the Earth.

Various sources of both radiation and particles exist on the Sun. Solar flares may release huge amounts of both radiation and particles. These events are unpredictable and sporadic. Solar radiation also varies over an approximately 27 day period, which is the rotational period of the Sun. The same source of radiation will face the Earth once every 27 days, so events tend to be somewhat repetitive.

Solar and galactic noise affect the reception of weak signals, while solar noise will also either affect radio propagation or act as a harbinger of changes in propagation patterns. Solar noise can be demonstrated by using an ordinary radio receiver and a directional antenna, preferably operating in the VHF/UHF regions of the spectrum. If the antenna is aimed at the Sun on the horizon at either sunset or sunrise a dramatic change in background noise will be noted as the Sun slides across the horizon.

Sunspots

A principal source of solar radiation, especially the periodic forms, is sunspots (Figure 1.4A). Sunspots can be as large as 100 000–150 000 km in diameter, and generally occur in clusters. The number of sunspots varies over a period of approximately 11 years, although the actual periods since 1750 (when records were first kept) have varied from 9 to 14 years (Fig. 1.4B). The sunspot number is reported daily as the statistically massaged **Zurich smoothed sunspot number**, or **Wolf number**. The number of sunspots greatly affects radio propagation via the ionosphere. The low was in the range of 60 (in 1907), while the high was about 200 (1958).

Another indicator of ionospheric propagation potential is the **solar flux index** (SFI). This measure is taken in the microwave region (wavelength of 10.2 cm, or 2.8 GHz), at 1700 U.T. Greenwich Mean Time in Ottawa, Canada. The SFI is reported by the National Institutes of Standards and Technology (NIST) radio stations WWV (Fort Collins, Colorado) and WWVH (Maui, Hawaii).

The ionosphere offers different properties that affect radio propagation at different times. Variations occur not only over the 11 year sunspot cycle but also diurnally and seasonally. Obviously, if the Sun affects propagation in a significant way, then differences between night-time and daytime, and between summer and winter, must cause variations in the propagation phenomena observed.

Ionospheric disturbances

Disturbances in the ionosphere can have a profound effect on radio communications – and most of them (but not all) are bad. In this section we will briefly examine some of the more common forms.

Sporadic E-layer

A reflective cloud of ionization sometimes appears in the E-layer of the ionosphere; this layer is sometimes called the E_s layer. It is believed that the E_s layer forms from the effects of wind shear between masses of air moving in opposite directions. This action appears to redistribute ions into a thin a layer that is radio-reflective.

Sporadic-E propagation is normally thought of as a VHF phenomenon, with most activity between 30 and 100 MHz, and decreasing activity up to about 100 MHz. However, about 25–50% of the time, sporadic-E propagation is possible on frequencies down to 10–15 MHz. Reception over paths of 2300–4200 km is possible in the 50 MHz region when sporadic-E propagation is present. In the northern hemisphere, the months of June and July are the most prevalent sporadic-E months. On most days when the sporadic-E phenomenon is present it lasts only a few hours.

Sudden ionospheric disturbances (SIDs)

The SID, or 'Dellinger fade,' mechanism occurs suddenly, and rarely gives any warning. Solar flares (Figure 1.5) are implicated in SIDs. The SID may last from a few minutes to many hours. It is believed that SIDs occur in correlation with solar flares or 'bright solar eruptions' that produce immense amounts of ultraviolet radiation that impinge the upper atmosphere. The SID causes a tremendous increase in D-layer ionization, which accounts for the radio propagation effects. The ionization is so intense that all receiver operators on the sunny side of the Earth experience profound loss of signal strength above about 3 MHz. It is not uncommon for receiver owners to think that their receivers are malfunctioning when this

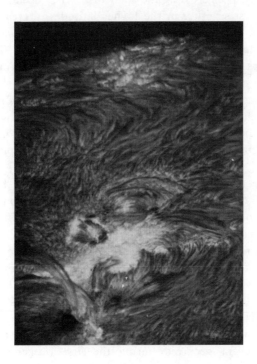

FIGURE 1.5

occurs. The sudden loss of signal by sunny-side receivers is called **Dellinger fade**. The SID is often accompanied by variations in terrestial electrical currents and magnetism levels.

An interesting anomaly is seen when SIDs occur. Although short-wave reception is disrupted, and may stay that way for awhile, distant very low-frequency (VLF) signals, especially in the 15–40 kHz region, experience a sudden increase in intensity. This is due to the fact that the SID event results in deep ionization way into the D-layer. This ionization increases absorption of HF signals. But the wavelength of VLF signals is close to the distance from the Earth's surface to the bottom of the D-layer, so that space acts like a gigantic 'waveguide' (as used in the transmission of microwaves) when the SID is present – thus propagating the VLF signal very efficiently.

Ionospheric storms

The ionospheric storm appears to be produced by an abnormally large rain of atomic particles in the upper atmosphere, and is often preceded by a SID 18–24 hours earlier. These storms tend to last from several hours to a week or more, and are often preceded by 2 days or so by an abnormally large collection of sunspots crossing the solar disk. They occur most frequently, and with greatest severity, in the higher latitudes, decreasing toward the Equator. When the ionospheric storm commences, short-wave radio signals may begin to flutter rapidly and then drop out altogether. The upper ionosphere becomes chaotic, turbulence increases, and the normal stratification into 'layers' or zones diminishes.

Radio propagation may come and go over the course of the storm, but it is mostly absent. The ionospheric storm, unlike the SID which affects the sunny side of the Earth, is worldwide. It is noted that the maximum usable frequency (MUF) and critical frequency tend to reduce rapidly as the storm commences.

An ionospheric disturbance observed over November 12–14, 1960 was preceded by about 30 minutes of extremely good, but abnormal propagation. At 15.00 hours EST, European stations were noted in North America with S9+ signal strengths in the 7000–7300 kHz region of the spectrum, which is an extremely rare occurrence. After about 30 minutes, the bottom dropped out, and even AM broadcast band skip (later that evening) was non-existent. At the time, WWV was broadcasting a 'W2' propagation prediction at 19 and 49 minutes after each hour. It was difficult to hear even the 5 MHz WWV frequency in the early hours of the disturbance, and it disappeared altogether for the next 48 hours. Of course, as luck would have it, that event occurred during the first weekend of the ARRL Sweepstakes ham radio operating contest that year.

GREAT CIRCLE PATHS

A great circle is a line between two points on the surface of a sphere that lies on a plane through the sphere's center. When translated to 'radio speak,' a great circle is the shortest path on the surface of the Earth between two points. Navigators and radio operators use the great circle for similar, but different reasons: the navigator in order to get from here to there, and the radio operator to get a transmission path from here to there.

The heading of a directional antenna is normally aimed at the receiving station along its great circle path. Unfortunately, many people do not understand the concept well enough, for they typically aim the antenna in the wrong direction. Radio waves do not travel along what appears to be the best route on a flat map. Instead they travel along the shortest distance on a real globe.

Long path versus short path

The Earth is a sphere (or more precisely, an 'oblique spheroid'), so from any given point to any other point there are two great circle paths: the **long path** (major arc) and **short path** (minor arc). In general, the best reception occurs along the short path. In addition, short-path propagation is more nearly 'textbook' compared with long-path reception. However, there are times when the long path is better, or is the only path that will deliver a signal to a specific location from the geographic location in question.

USING THE IONOSPHERE

The refraction of HF and some medium-wave radio signals back to Earth via the ionosphere gives rise to intercontinental HF radio communications. This phenomenon becomes possible during daylight hours, and for a while after sunset when the ionosphere is ionized. Figure 1.6 reiterates the mechanism of long-distance skip communications. The transmitter is located at point T, while receiving stations are located at sites R1 and R2. Signals 1 and 2 are not refracted sufficiently to be returned to Earth, so they are lost in space. Signal 3, however, is refracted enough to return to Earth, so it is heard at station R1. The skip distance for signal 3 is the distance from T to R1. At points between T and R1, signal 3 is inaudible, except within ground wave distance of the transmitter site (T). This is the reason why two stations 50 km apart hear each other only weakly, or not at all, while both stations can communicate with a third station 3000 km away. In American amateur radio circles it is common for South American stations to relay between two US stations only a few kilometers apart.

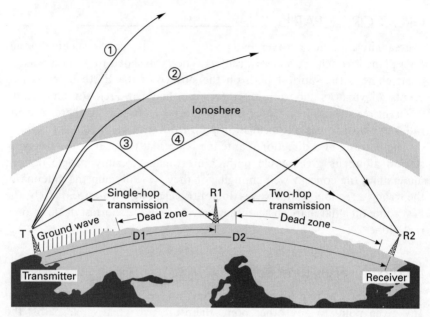

FIGURE 1.6

Multi-hop skip is responsible for the reception of signal 3 at site R2. The signal *reflects* (not refracts) from the surface at R1, and is retransmitted into the ionosphere, where it is again refracted back to Earth.

The location where skip signals are received (at different distances) depends partially upon the **angle of radiation** of the transmitting antenna. A high angle of radiation causes a shorter skip zone, while a lower angle of radiation results in a longer skip zone. Communication between any particular locations on any given frequency requires adjustment of the antenna radiation angle. Some international short-wave stations have multiple antennas with different radiation angles to ensure that the correct skip distances are available.

SUPER-REFRACTION AND SUBREFRACTION

At VHF frequencies and well up into the microwave bands there are special propagation modes called *super-refraction* and *subrefraction*. Depending upon the temperature gradient and humidity, the propagation may not be straight line. At issue is something called the *K-factor* of the wave. Figure 1.7 shows these modalities. The straight line case has a K-factor of one, and is the reference. If super-refraction occurs, then the value of K is greater than one, and if subrefraction occurs it is less than one.

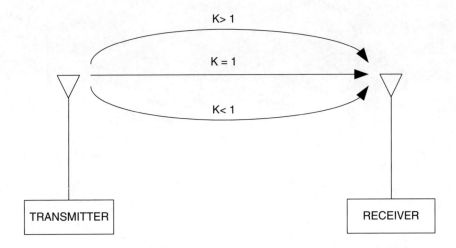

FIGURE 1.7

Figure 1.8 shows a case where super-refraction occurs. The value of K can be substantially above one in cases where a hot body of land occurs next to a relatively cool body of water. This occurs off Baja, California, the Persian Gulf and Indian Ocean, parts of Australia and the North African coast. In those areas, there may be substantial amounts of super-refraction occurring, making directional antennas point in the wrong direction.

Figure 1.9 shows a case of subrefraction. In this case, there is a relatively cold land mass next to relatively warm seas. In the Arctic and Antarctic this situation exists. The K factor will be substantially less than one in these cases. In fact, the signal may be lost to terrestrial communications after only a relatively short distance.

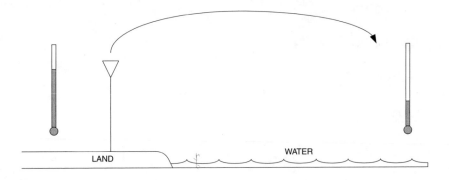

FIGURE 1.8

RADIO SIGNALS ON THE MOVE   **17**

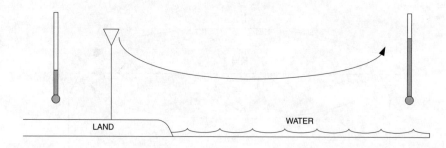

FIGURE 1.9

CHAPTER 2

Antenna basics

Before looking at the various antennas we need to look at some of the basics of antenna systems. In this chapter you will learn some of these basics. And while they will not make you a red-hot professional antenna engineer they will set you up well enough to understand this book and others on amateur and hobbyist antennas. We will look at the matter of antenna radiation, antenna patterns, the symbols used to represent antennas, voltage standing wave ratio (VSWR), impedance, and various methods suitable for constructing wire antennas in the high-frequency (HF) and very high-frequency (VHF) regions of the spectrum.

ANTENNA SYSTEM SYMBOLS

Figures 2.1 and 2.2 show the various symbols used to represent antennas and grounds. The reason why there are so many variants is that there are differences from country to country, as well as different practices within any one country (especially between technical publishers). As for antenna symbols, I see the symbol in Figure 2.1C more often in the USA, but Figure 2.1B comes in for a close second. The supposedly correct symbol (endorsed by a professional society drawing standards committee) is that of Figure 2.1A – but it is only occasionally seen in the USA.

The situation for grounds is a little different because some differences reflect different forms of ground (although some of the differences also represent national or publisher differences). The ground in Figure 2.2A is usually found representing a true earth ground, i.e. the wire is connected to a

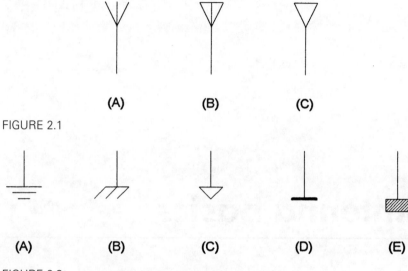

FIGURE 2.1

FIGURE 2.2

rod driven into the earth. The variant of Figure 2.2B usually represents a chassis ground inside a piece of equipment. The symbol in Figure 2.2C has two uses. One is to represent a common grounding point for different signals or different pieces of equipment. The other use is exactly the opposite: the triangle ground symbol often represents an **isolated ground** that has no direct electrical connection to the rest of the circuit, or with the earth. You will see this usage in medical devices. The grounds of Figures 2.2D and 2.2E are found mostly outside the USA.

From the time of Hertz and Marconi to the present, one thing has remained constant in wireless communications: radio waves travel, as if by magic, from a transmitting antenna to a receiving antenna (Figure 2.3). Whether the two antennas are across the garden from each other, across continents and oceans, or on the Earth and the Moon, if there is not a transmitting antenna and at least one receiving antenna in the system then no communications can take place.

At one time, physicists believed that there must be some invisible medium for carrying the radio signal. But we now know that no such medium exists, yet radio waves travel even in outer space. Being electromagnetic waves, radio signals need no medium in order to propagate. If radio signals traveled only in the Earth's atmosphere, then we could make some guesses about a medium for carrying the wave, but space communications demonstrates that the atmosphere is not necessary (although it does affect radio signal propagation).

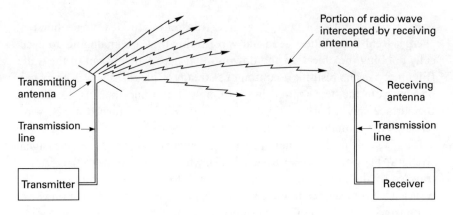

FIGURE 2.3

Although there is no medium in which radio waves travel, it is useful to look at water waves for an analogy (even though imperfect). In Figure 2.4 we see what happens when an object is dropped into a pool of water. A displacement takes place, which forms a leading wave that pushes out in concentric circles from the impact point. The situation in Figure 2.4 repre-

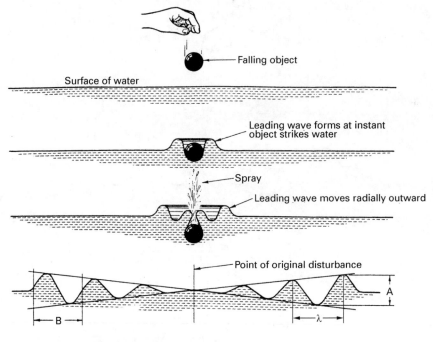

Notes: A, amplitude of leading wave.
B, corresponds to 1 cycle of oscillation.

FIGURE 2.4

ANTENNA BASICS **21**

sents a single pulse of energy, as if a transmitter fired a single burst of energy. Real transmitters send out wave trains that are analogous to cyclically bobbing the object up and down so that it goes in and out of the water (Figure 2.5). The result is a continuous stream of identical waves propagating out from the 'transmitter' impact point. If another object is floating on the surface, say a cork or toy boat, then it will be perturbed as the wave passes. This is analogous to the receiver antenna.

The waves have an **amplitude** ('A'), which corresponds to the signal strength. They also have a **wavelength** (λ), which corresponds to the distance traveled by the wave in one complete up-and-down cycle. In radio work, the wavelength is measured in **meters** (m), except in the microwave region where **centimeters** (cm) and **millimeters** (mm) make more sense. Wavelength can be measured at any pair of points on the wave that are identical: two peaks, two troughs, two zero crossings, as convenient in any specific case.

The number of cycles that pass a given point every second is the **frequency** of the wave. The classic measure of frequency was cycles per second (cps or c/s), but that was changed in 1960 by international consensus to the hertz (Hz), in honor of Heinrich Hertz. But since 1 Hz = 1 cps, there is no

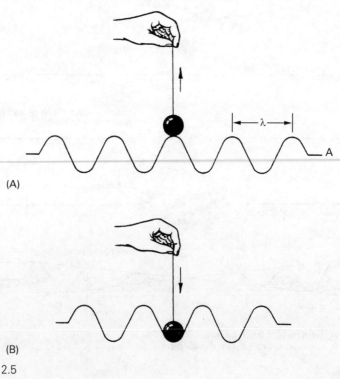

FIGURE 2.5

practical difference. The hertz is too small a unit for most radio work (although many of our equations are written in terms of hertz). For radio work the **kilohertz** (kHz) and **megahertz** (MHz) are used: 1 kHz = 1000 Hz, and 1 MHz = 1 000 000 Hz. Thus, a short-wave frequency of 9.75 MHz is 9750 kHz and 9 750 000 Hz.

Wavelength and frequency are related to each other. The wavelength is the reciprocal of frequency, and vice versa, through the velocity constant. In free space, the velocity constant is the speed of light (c), or about 300 000 000 m/s. This is the reason why you often see '300' or its submultiples (150 and 75) in equations. When the frequency is specified in megahertz, then 300 000 000 becomes 300 for one wavelength. The half-wavelength constant is 150, and the quarter-wavelength constant is 75. The relationship is

$$\lambda_{meters} = \frac{300}{F_{MHz}}$$

INVERSE SQUARE LAW

When radio waves travel they become weaker by a relationship called the **inverse square law**. This means that the strength is inversely proportional to the square of the distance traveled ($1/D^2$). Figure 2.6 shows how this works using the analogy of a candle. If the candle projects a distance r, all of the light energy falls onto square 'A'. At twice the distance ($2r$) the light spreads out and covers four times the area (square 'B'). The total amount of light energy is the same, but the energy per unit of area is reduced to one-fourth of the energy that was measured at 'A.' This means that a radio signal gets weaker very rapidly as the distance from the transmitter increases, requiring ever more sensitive receivers and better antennas.

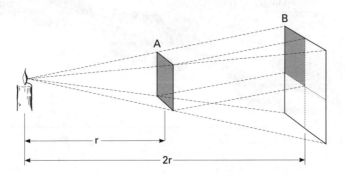

FIGURE 2.6

THE ELECTROMAGNETIC WAVE

The electromagnetic (EM) wave propagating in space is what we know as a 'radio signal.' The EM wave is launched when an electrical current oscillates in the transmitting antenna (Figure 2.7). Because moving electrical currents possess both electrical (E) and magnetic (H) fields, the electromagnetic wave launched into space has alternating E-field and H-field components. These fields are **transverse** (meaning they travel in the same direction) and **orthogonal** (meaning the E- and H-fields are at right angles to each other). When the EM wave intercepts the receiver antenna, it sets up a copy of the original oscillating currents in the antenna, and these currents are what the receiver circuitry senses.

The orthogonal E- and H-fields are important to the antenna designer. If you could look directly at an oncoming EM wave, you would see a plane front advancing from the transmitting antenna. If you had some magical dye that would render the E-field and H-field line of force vectors visible to the naked eye, then you would see the E-field pointing in one direction, and the H-field in a direction 90° away (Figure 2.8).

The **polarization** of the signal is the *direction of the E-field vector*. In Figure 2.8 the polarization is vertical because the electric field vector is up and down. If the E-field vector were side-to-side, then the polarization would be horizontal. One way to tell which polarization an antenna produces when it transmits, or is most sensitive to when it receives, is to note the direction of the radiator element. If the radiator element is vertical, i.e. perpendicular to the Earth's surface, then it is vertically polarized. But if

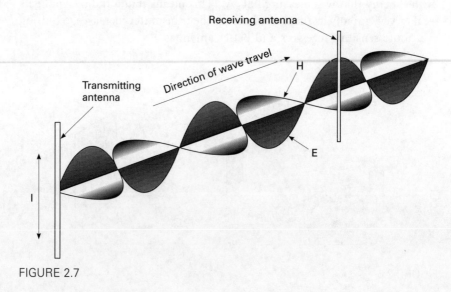

FIGURE 2.7

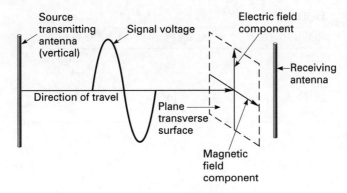

FIGURE 2.8

the radiator element is horizontal with respect to the Earth's surface, then it is horizontally polarized. Figure 2.9 shows these relationships. In Figure 2.9, two dipole receiver antennas are shown, one is vertically polarized (VD) and the other is horizontally polarized (HD). In Figure 2.9A, the arriving signal is vertically polarized. Because the E-field vectors lines are vertical, they cut across more of the VD antenna than the HD, producing a considerably larger signal level. The opposite is seen in Figure 2.9B. Here the E-field is horizontally polarized, so it is the HD antenna that receives the most signal. The signal level difference can be as much as 20 dB, which represents a 10-fold decrease in signal strength if the wrong antenna is used.

DECIBELS (dB)

In the section above the term **decibel** (symbol 'dB') was introduced. The decibel is a unit of measure of the ratio of two signals: two voltages, two currents, or two powers. The equations for decibels take the logarithm of the ratio, and multiply it by a constant (10 for powers and 20 for voltages or currents). The use of decibel notation makes it possible to use **ratios**, such as found in gains and losses in electronic circuits, but use only addition and subtraction arithmetic. It is not necessary to be able to calculate decibels, but you should know that +dB represents a gain, and −dB represents a loss. The term '0 dB' means that the ratio of the two signals is 1:1 (neither gain nor loss). Some common ratios encountered in radio work include those listed in Table 2.1.

You can see that doubling a signal strength results in a +3 dB gain, while halving it produces a −3 dB loss. To put these figures into perspective, most S-meters on receivers use a scaling factor of 6 dB per S-unit (some use 3 dB per S-unit). Receiver designers tell us that a signal-to-noise ratio of 10 dB is necessary for 'comfortable listening,' while a signal-to-noise ratio of 3 dB is

TABLE 2.1

Ratio	Power (dB)	Voltage (dB)
1:1	0	0
2:1	+3	+6
1:2	−3	−6
10:1	+10	+20
1:10	−10	−20

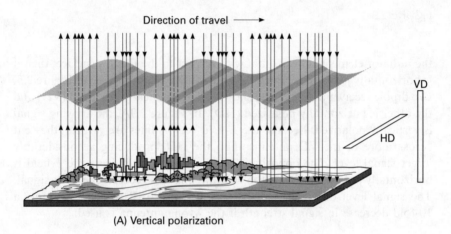

FIGURE 2.9

needed for barely perceptible but reliable communication for a listener who tries hard to hear what is being said.

LAW OF RECIPROCITY

Radio antennas obey a kind of **law of reciprocity**, i.e. they work the same on transmit as they do on receive. If an antenna has a certain gain and directivity on transmit, then that exact same pattern is seen in the receive mode. Similarly, the feedpoint impedance, element lengths, spacings, and other issues are the same for both modes. An implication of this law is that receiver owners are able to translate the theory from discussions of transmitting antennas to their own needs (and vice versa).

Reciprocity does not necessarily mean that the same antennas are the best selection for both transmit and receive. For example, the use of gain on a transmit antenna is to increase the apparent signal power level at a distant point in a particular direction. Alternatively, some licensing authorities use antenna directivity to protect the coverage areas of relatively nearby stations that share the same or adjacent frequencies. To the receiver operator, there may be a reason to want the gain of the antenna to boost the received signal power to a useful level. There is, however, often a powerful argument for aiming an antenna such that the main sensitivity is not in the direction of the desired station. Instead, the receiver antenna owner may wish to position a null, i.e. the least sensitive aspect of the antenna, in the direction of an offending station in order to reduce its effect. The issue is, after all is said and done, the signal-to-noise ratio.

Another restriction is that there are several antennas that are fit for receive use, especially those for difficult installation situations, but are either technically unsuited or unsafe for transmitters except at the lowest of power levels. The issue might be impedance, VSWR, voltage arcing, or some other undesirable factor that occurs in transmit situations (even at moderate power levels such as 50–100 W).

ANTENNA TYPES

Most antennas fall into either of two broad categories: **Marconi** or **Hertzian**. The Marconi antennas (Figure 2.10) are considered unbalanced with respect to ground because one side of the signal source (or receiver input load) is grounded. The Marconi antenna can be either vertical (Figure 2.10A) or at some angle (Figure 2.10B), including horizontal. The name derives from the types of antenna used by Guglielmo Marconi in his early radio experiments. In his famous 1903 transatlantic communication, ol' 'Gug' used a single wire tethered to a kite aloft above the Newfoundland coast.

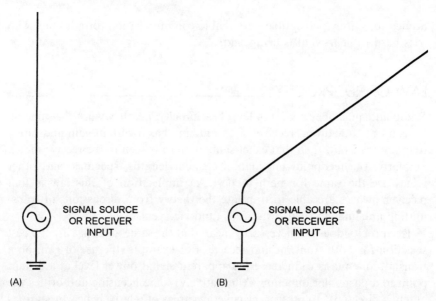

FIGURE 2.10

The Hertzian antennas (Figure 2.11) are balanced with respect to the ground, i.e. neither side of the oscillator or receiver load is grounded. Dipoles fall into this category. Both Marconi and Hertzian antennas can be either vertically or horizontally polarized. However, most Marconi antennas are designed for vertical polarization and most Hertzian antennas for horizontal polarization. The long wire and vertical dipoles are obvious exceptions to that rule, however.

RECEIVER–ANTENNA INTERACTIONS

The antenna and the receiver (or the antenna and transmitter on the other end) form a system that must be used together. Either alone is not too useful. Figure 2.12 shows an antenna connected to a receiver through a transmission line. For our practical purpose here, the receiver looks to the

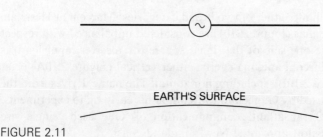

FIGURE 2.11

28 ANTENNA TOOLKIT

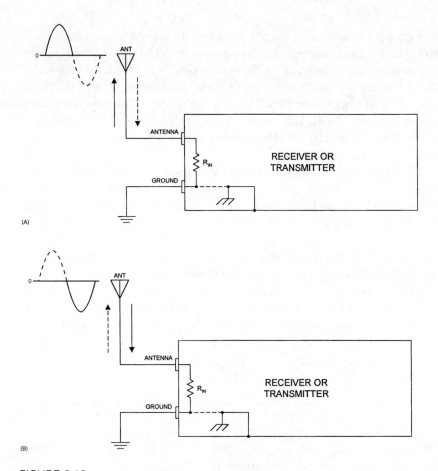

FIGURE 2.12

antenna and transmission line like a load resistor, R_{IN}. On the positive half-cycle (Figure 2.12A), the passing wave creates a current in the antenna–receiver system that flows in one direction (shown here as 'up'). When the oscillation reverses and becomes negative, the direction of current flow in the antenna–receiver reverses (Figure 2.12B). This is the mechanism by which the oscillating electromagnetic wave reproduces a signal in the input of the receiver – the signal that is then amplified and demodulated to recover whatever passes for 'intelligence' riding on the signal.

STANDING WAVE RATIO

The issue of the **standing wave ratio** (SWR) is of constant interest to radio enthusiasts. Some of the heat and smoke on this matter is well justified. In other cases, the perceived problems are not real.

Figure 2.13 shows how the SWR comes into play in an antenna system. In Figure 2.13A, a single cycle of a signal is launched down a transmission line (it is called the '**incident**' or '**forward**' wave). When it reaches the end of the line, if it is not totally absorbed by a load resistor or antenna, then it (or part of it) will be reflected back toward the source. This **reflected** wave is shown in Figure 2.13B. The incident and reflected waves are both examples of **travelling waves**. The reflected wave represents power that is lost, and can cause other problems as well.

The situation in Figures 2.13A and 2.13B represent a single-cycle pulse launched down a transmission line. In a real radio system, the oscillations of the incident wave are constant (Figure 2.13C). When this situation occurs, then the reflected waves will interfere with following incident waves. At any given point, the amplitude of the wave is the algebraic sum of the interfering incident and reflected signals. The resultant caused by the interference of the incident and reflected waves is called a **standing wave**.

Figure 2.14 shows what happens when continuous incident and reflected waves coexist on the same transmission line. In the case of Figure 2.14A, the two waves coincide, with the resultant as shown. The waves begin to move

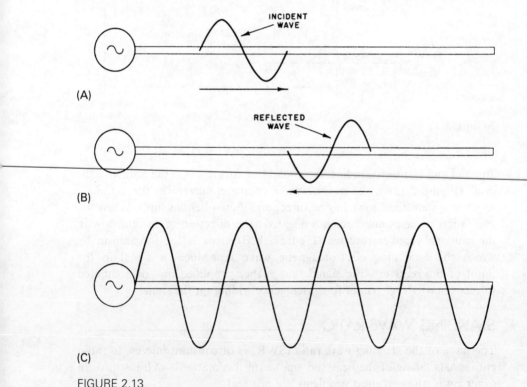

FIGURE 2.13

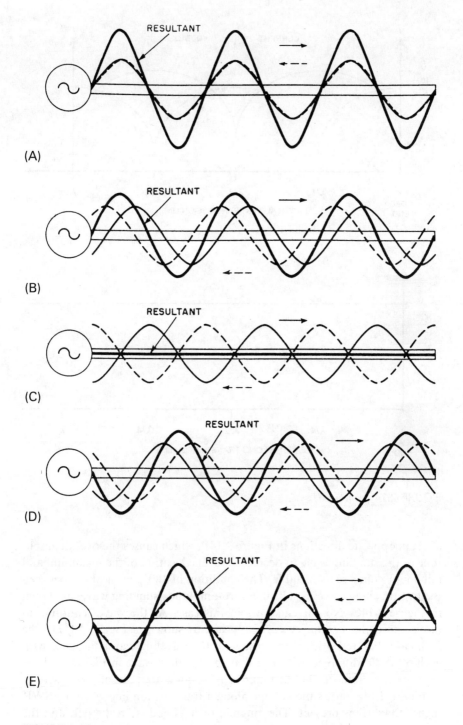

FIGURE 2.14

ANTENNA BASICS

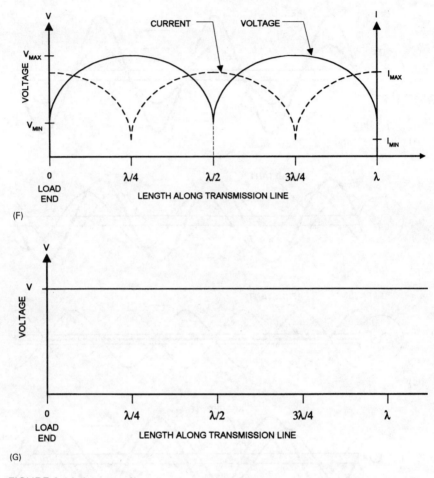

FIGURE 2.14 (continued)

apart in opposite directions in Figure 2.14B, which causes the overall amplitude of the standing wave to decrease, but the location of the maximum and minimum points remain stable. The condition of 180° out of phase between the two traveling waves results in a zero-amplitude standing wave, as shown in Figure 2.14C. No current flows in this case. As the waves continue to move apart, the standing wave reappears, as shown in Figure 2.14D. The final case, Figure 2.14E, is the case with the traveling waves in-phase with each other (0° difference). Notice that the standing wave in Figure 2.14E is the same amplitude as that of Figure 2.14A, but is 180° out of phase with it.

Figure 2.14F shows the voltage along a transmission line when a VSWR higher than 1:1 is present. The antenna or load end is on the left, and the lengths along the transmission line are plotted in terms of wavelength. In the situation shown, the voltage is a minimum at the antenna end (marked '0'). It rises to a peak or 'antinode' at a quarter wavelength, and then drops back to a minimum ('node') at the half-wavelength ($\lambda/2$) point. The voltage then rises again to a peak at $3\lambda/4$, falling back to minimum at 1λ. The current rises and falls in a similar manner, but the nodes and antinodes are offset from the voltage by 90° (quarter wavelength).

Note that the voltage is at a minimum and the current at a maximum at the same points that are integer multiples of a half wavelength. This situation has the effect of making it advisable to measure the VSWR at the transmitter end only when integer multiples of an electrical half wavelength of transmission line (note: the most valid measurement is made at the interface of the transmission line and antenna feedpoint).

The plot shown in Figure 2.14G is the same line system under the situation where the antenna impedance and transmission line impedance are matched. For obvious reasons, this line is said to be 'flat.'

Measures of SWR

There are several different methods for measuring the SWR of an antenna transmission line system. For example, one could measure the incident and reflected signal voltages along the line, producing a result of $VSWR = (V_i + V_r)/(V_i - V_r)$. The method of Figures 2.14F and 2.14G can be used by comparing the maximum and minimum voltages: $VSWR = V_{MAX}/V_{MIN}$. We can also measure the forward and reverse power levels to find the VSWR. A simple way to predict the VSWR is to compare the antenna feedpoint resistive impedance (Z_L) to the transmission line characteristic impedance (Z_0). The value of the VSWR is found from either

$$VSWR = \frac{Z_0}{Z_L} \qquad (Z_0 > Z_L)$$

or

$$VSWR = \frac{Z_L}{Z_0} \qquad (Z_L > Z_0)$$

If, for example, we measure the antenna feedpoint impedance as being 25 ohms, and the antenna transmission line is 52 ohm coaxial cable, then the $VSWR = 52/25 = 2.08:1$.

We can find the resonant frequency of an antenna by finding the frequency at which the VSWR is a minimum. But that point may not be a VSWR of 1:1 unless the resistive component of the antenna feedpoint impedance is the same as the transmission line impedance, and the complex portion of the impedance is zero. To create this situation one might need an impedance matching device (see Chapter 12).

AN OLD MYTH REVIVED?

There are a number of myths that are widely held among radio communications hobbyists – and amateur radio is no less infested with some of these myths than others (CB, for example). Twenty-five years ago I worked in a CB shop in Virginia, and we kept hearing one old saw over and over again: you can 'cut your coax to reduce the VSWR to 1' (actually, they meant '1:1' but routinely called it '1'). Hoards of CBers have 'cut the coax' and watched the VSWR reduce to 1:1, so they cannot be talked out of the error. I even know of one shop that kept 30 cm lengths of coaxial cable, with connectors on both ends, so they could insert them into the line at the transmitter in order to find the correct length that would reduce the VSWR to 1:1. What actually happens in that case is a measurement artifact that makes it appear to be true.

Of course, hams are superior to CBers and so do not believe that error, right? I would like to think so; but having been in both the CB and the amateur worlds, and 'Elmered' (mentored) more than a few CBers studying for amateur licenses, I have to admit that at least as many amateurs believe the 'cut the coax' error as CBers (sorry, fellows, but that's my observation).

The only really proper way to reduce the VSWR to 1:1 is to tune the antenna to resonance and then match the impedance. For a center-fed half-wavelength dipole, or a bottom-fed quarter-wavelength vertical, the proper way to resonate the antenna is to adjust its length to the correct point. The formulas in the books and magazines only give approximate lengths – the real length is found from experimentation on the particular antenna after it is installed. Even commercial antennas are adjusted this way. On certain CB mobile antennas, for example, this trick is done by raising (or lowering) the radiator while watching the VSWR meter. On amateur antennas similar tuning procedures are used.

Even when the resonant point is found, the feedpoint impedance may not be a good match to the transmission line. A VSWR will result in that case. The impedance matching should be done between the far end of the transmission line (i.e. away from the receiver or rig) at the feedpoint of the antenna. Antenna tuners intended for strictly coaxial cable are little more than line flatteners. They do not really 'tune' the antenna, but rather they reduce the VSWR looking into the transmission line so that the transmitter will work properly. If the antenna tuner is not a high-pass filter (as some are), then it will also provide some harmonic attenuation.

An approach used by many amateurs (including myself) is to connect an antenna-matching unit (tuner) at the output of the transmitter. For my Kenwood TS-430, I use either a Heath SA-2060A or an MFJ Differential Tuner to 'tune-out' the VSWR presented by my Hustler 4BTV and 23 m of coaxial cable. But I do not even pretend to be tuning the antenna. The

TS-430 is a solid-state rig, and the finals are, therefore, not terribly tolerant of VSWR, and will shut down with a high VSWR. The purpose of the antenna tuner is to reduce the VSWR seen by the transmitter – ignoring the actual antenna mismatch on the roof. The tuner also serves to reduce harmonics further, thereby helping to prevent television interference.

The best form of antenna tuner is one that both reduces the VSWR (for the benefit of the transmitter), and also resonates to the antenna frequency, preventing harmonics from getting out (a little secret is that many 'line flattener' antenna tuning units are actually variable high-pass filters, and must be used with a low-pass filter ahead of them if spurious signals are to be kept in abeyance).

Should you even worry about VSWR on a system? Or, more correctly stated, given that a 1:1 VSWR often requires a herculean effort to achieve, at what point do you declare the battle won and send the troops home? Some of the issues are:

- Transmitter heating of the transmission line due to power losses.
- Reduction of power from solid-state transmitters due to SWR shutdown circuitry.
- At kilowatt levels there may be excessive radio frequency voltage at nodes, which could lead to transmission line shorting.
- Loss of receiver sensitivity. The feedline loss is added directly to the receiver noise figure, so may degrade the matched noise figure of the receiver. This problem is especially severe in the VHF/UHF scanner bands.

Where these problems result in unacceptable performance, work hard to match the antenna to the feedline, and the feedline to the transmitter or receiver. As one of my correspondents said: 'The question then is to what extent does the problem impair the actually ability to use the receiver or transmitter successfully?' The same reader provided the guidelines below:

(1) The actual SWR on a feedline is 1:1 only if the load (e.g. an antenna) connected to the line is equal to the characteristic impedance of the line. Adjusting an antenna to resonance will improve the SWR only if the impedance of the antenna at resonance is closer to the impedance of the feedline than before it was adjusted. If the impedance of the antenna at resonance is not equal to the impedance of the feedline, you will never get the SWR down to 1:1.

(2) The only possible places to measure the SWR with consistent accuracy are:
 (a) at the load
 (b) at a distance of 0.5 electrical wavelength (accounting for the velocity factor of the line), or integer multiples thereof, from the load.

The latter is true because a half-wavelength long piece of feedline 'repeats' the impedance of whatever is connected to its far end; it is almost as if you are measuring right up at the feedpoint. As a consequence, many professional antenna installers go to pains to make the transmission line $\lambda/2$ or an integer multiple of $\lambda/2$. The VSWR and impedance looking into the line will reflect the situation at the feedpoint. Be aware that the velocity factor of the transmission line reduces the physical length required to make an electrical half wavelength.

(3) An easy way to check if your SWR measurements are being accurately made is as follows: Add a piece ($\lambda/8$ to $\lambda/4$ wavelength) of *identical* line to your feedline. Repeat the SWR measurement. If it is not the same, your SWR measurement is not accurate, i.e. you are not measuring the actual SWR.

(4) Know that a few things can upset SWR measurements, e.g. currents flowing on the outside of a coaxial feedline (a common situation). This can be caused by not using a BALUN when feeding a balanced antenna (e.g. a symmetric dipole) with an unbalanced (coaxial) feedline. It can also be induced by antenna currents onto a nearby coaxial feedline.

ANTENNA GAIN, DIRECTIVITY, AND RADIATION PATTERNS

Antennas do not radiate or receive uniformingly in all directions (and, indeed, you usually do not want them to!). The interrelated concepts of **antenna gain** and **directivity** are of much concern to antenna builders. Gain refers to the fact that certain antennas cause a signal to seem to have more power than it actually can claim. The gain of the antenna can be expressed as either a multiple (e.g. a twofold increase in apparent power) or in decibel notation (e.g. a 3 dB increase). Both methods of expressing gain reflect the fact that a gain antenna seems to produce a signal that is stronger than the signal produced by some comparison antenna (e.g. a dipole). The apparent power produced by the antenna and transmitter combination is called the **effective radiated power** (ERP), and is the product of the transmitter power and the antenna gain. For example, if the antenna gain is 7 dB, the power is increased by a factor of five times. The ERP of a 100 W transmitter is therefore $5 \times 100\,W = 500\,W$.

On a receive-only antenna, the signal picked up by a gain antenna is louder than the same signal picked up on a non-gain antenna or the comparison antenna. Because of reciprocity, the gain is the same for both receive and transmit.

Directivity refers to the fact that the radiation or reception direction is not constant. Certain directions are favored and others rejected. Indeed, this is how the antenna gets gain. After all, if there is an apparent increase in the signal power, there must be some phenomenon to account for it – after all, antennas do not produce power. Figure 2.15A shows how the gain and directivity are formed. Consider an **isotropic radiator** source located at point 'A' in Figure 2.15A. An isotropic radiator is a perfectly spherical

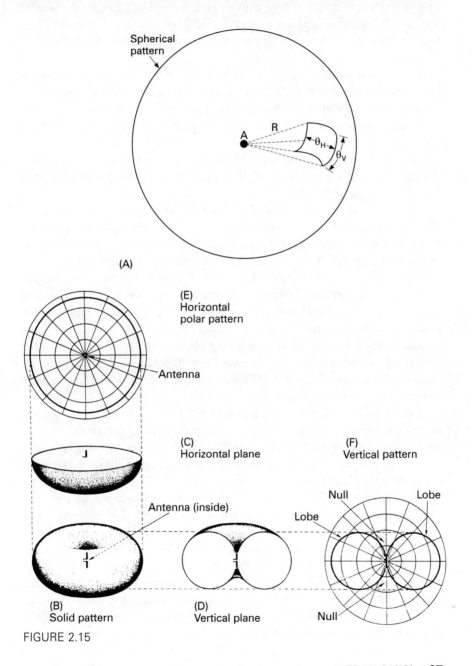

FIGURE 2.15

point source. The signal radiates outward from it in a spherical pattern. At some distance R we can examine the surface of the sphere. The entire output power of the transmitter is uniformly distributed over the entire surface of the sphere. But if the antenna is directive, then all of the power is focussed onto a small area of the sphere with a vertical angle of θ_V and a horizontal area of θ_H. The same amount of energy exists, but now it is focussed onto a smaller area, which makes it appear as if the actual power level was a lot higher in that direction.

The **antenna pattern** is a graphical way of showing the relative radiation in different directions (i.e. directivity). The most commonly seen pattern is the **azimuthal** or **polar** plot. That pattern is measured in the horizontal plane at all angles around the antenna, and is plotted graphically as if seen from above. It is a 'bird's eye view' of the antenna radiation characteristic. But antenna radiation patterns are actually three-dimensional, and the azimuthal plot is merely a slice of that three-dimensional pattern, as seen from but one perspective.

Figures 2.15B through 2.15F show how antenna patterns are developed. The antenna in this case is a vertically polarized Hertzian radiator such as a half-wavelength dipole. It radiates poorly off the ends, and very well at angles perpendicular to the wire. The solid-pattern of Figure 2.15B shows the three-dimensional view of what the radiation looks like. To find the azimuthal and elevation patterns for this antenna, we take slices out of the solid (Figures 2.15C and 2.15D), and plot the results (Figures 2.15E and 2.15F). Because the radiator is vertical (and perfect, I might add), it radiates equally well in all horizontal directions (see Figure 2.15E), so the azimuthal or horizontal plot is termed **omnidirectional**. The elevation or vertical plane view (Figure 2.15F) has to take into account the minima found off the ends of the radiator, so plots as a figure '8.'

If the vertical radiator of Figures 2.15B–2.15F is rotated 90° to form a horizontal dipole, then the radiation patterns remain the same, but are reversed. The omnidirectional pattern of Figure 2.15E becomes the elevation or vertical extent, while the figure '8' pattern of Figure 2.15F is the horizontal or azimuthal extent.

The patterns of Figures 2.15B–2.15F make a rather bold and unwarranted assumption: the antenna is perfect and is installed somewhere in free space. Since most antennas are not in the zone midway between the Earth and Mars, one has to account for the effects of being close to the Earth's surface. Figure 2.16 shows the approximate patterns for horizontally polarized half-wavelength dipoles close to the ground. In Figure 2.16 we see the azimuthal pattern. It is the figure '8,' but is pinched a bit. Notice that the maxima are perpendicular to the wire, and the minima (or 'nulls') are off the ends. One use for this type of pattern is to position the antenna so that it

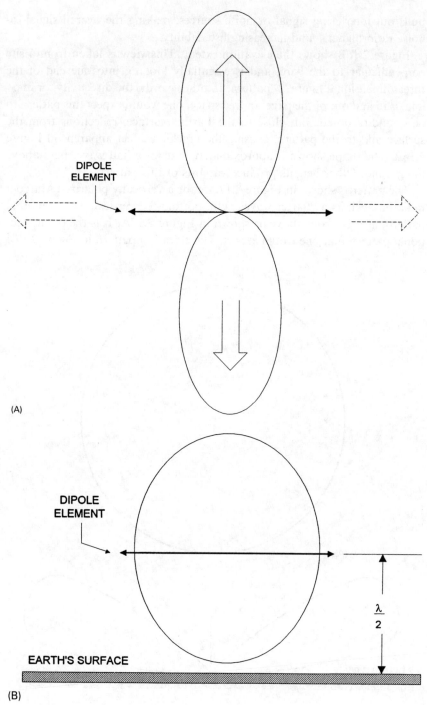

(A)

(B)

FIGURE 2.16

nulls out interfering signal or noise sources, making the overall signal-to-noise ratio higher – and increasing listenability.

Figure 2.16B shows the elevation extent. This view is taken from a site perpendicular to the wire, and is essentially looking into one end of the three-dimensional figure '8' pattern. In other words, the directivity, or maxima, is in and out of the page. In free space, one would expect this pattern to be omnidirectional. But close to the Earth's surface, reflections from the surface add to the pattern, causing the pinching effect apparent in Figure 2.16B. The shape shown is approximately that for a half wavelength above the ground. Other heights produce variants of this pattern.

The patterns shown in Figure 2.17 are for a vertically polarized Marconi radiator, such as a quarter-wavelength, bottom-fed vertical. The azimuthal pattern, as seen from above, is shown in Figure 2.17A. It is the omnidirectional pattern that one would expect. The elevation pattern in Figure 2.17B

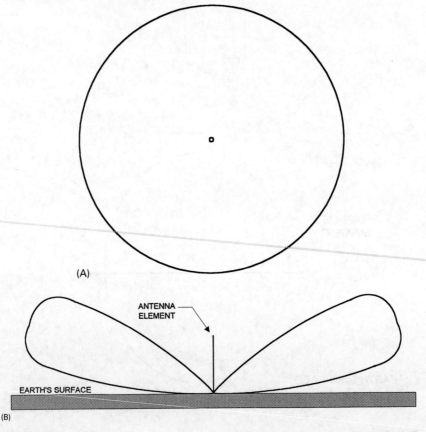

FIGURE 2.17

represents a vertical slice taken from the 'doughnut' seen earlier. Note that the lobes are elevated, rather than being exactly on the horizon. The angle of these lobes is the **angle of radiation**, and affects the distance that an antenna will transmit to or receive from. A low angle of radiation is usually prescribed for long-distance 'DX,' while a high angle of radiation produces skip to nearer regions.

Figure 2.18 shows the conventions for specifying antenna patterns. The antenna is a dipole, with both relative electric field strength and relative powers plotted (both types of plot are done). The 0–270° line is along the length of the dipole, while the 90–180° line is perpendicular to the dipole. The **beamwidth** of the antenna, which is specified in degrees, is measured as the angle between the **half-power points**, i.e. the points where the voltage field drops off to 70.7% of the maximum (along the 90° line), or to the 0.5 power point. These points are also called the '−3 dB' points. Recall that −3 dB represents a power ratio of 1:2.

SOME ANTENNA CONSTRUCTION HINTS

This section will help you think up some good ideas on how to erect the antenna. First, though, let us talk a little bit about antenna safety. It is difficult to say too much about this topic. Indeed, I mention it elsewhere in this book as well as here. Wherever it seems prudent, I warn readers that

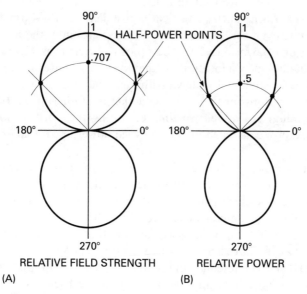

FIGURE 2.18

antenna erection can be dangerous if you do not follow certain precautions. Do it right, and the risk is mitigated and the job can be quite safe.

Safety first!

Before dealing with the radio and performance issues, let us first deal with safety matters – you do not want to be hurt either during installation, or during the next wind storm. Two problems present themselves: reliable mechanical installation and electrical safety.

Electrical safety note. Every year we read sad news in the magazines of a colleague being electrocuted while installing, or working on, an antenna. In all of these tragic cases, the antenna somehow came into contact with the electrical power lines. Keep in mind one dictum and make it an absolute:

There is never a time or situation when it is safe to let an antenna contact the electrical power lines! None. Ever. BELIEVE THIS NO MATTER WHAT ANYONE SAYS TO THE CONTRARY.

This advice includes dipoles and long wires 'thrown over' supposedly insulated power mains lines, as well as antennas built from aluminum tubing. The excuse that the lines are insulated is nonsense. Old insulation crumbles on contact with even a thin wire antenna. Do not do it! The operant word is **never**.

Consider a typical scenario involving a four-band trap vertical antenna made of tubing. It will be 5–8 m tall (judging from adverts in magazines), and will be mounted on a roof, or mast, 4–10 m off the ground. At my home in Virginia, a 7.6 m tall trap vertical is installed atop a 4.6 m telescoping television antenna mast. The total height above ground is the sum of the two heights: 7.6 m + 4.6 m = 12.2 m. The tip of the vertical is 12.2 m above the ground. I had to select a location, on the side of my house, at which a 12.2 m aluminum pole could fall safely. Although that requirement limited the selection of locations for the antenna, neither my father-in-law (who helped install the thing) nor myself was injured during the work session. Neither will a wind storm cause a shorted or downed power line if that antenna falls over.

In some jurisdictions, there might be legal limitations on antenna location. For example, some local governments in the USA have a requirement that the antenna be able to fall over and land entirely on your own property. Before installing the antenna, check local building codes.

When installing a vertical, especially one that is not ground mounted, make sure that you have help. It takes at least two people to safely install the standard HF vertical antenna, and more may be needed for especially large models. If you are alone, then go and find some friends to lend an arm or two. Wrenched backs, smashed antenna (and house parts), and other calamities simply do not happen as often to a well-organized work party that has a sufficient number to do the job safely.

Mechanical integrity

The second issue in installing antennas is old-fashioned mechanical integrity. Two problems are seen. First, you must comply with local building regulations and inspections. Even though the courts in the USA seem to forbid local governments from prohibiting amateur radio activity (on grounds that it is Federal prerogative), local governments in the USA and eslewhere have a justifiable interest, and absolute right, to impose reasonable engineering standards on the mechanical installation of radio antennas. The second issue is that it is in your own best interest to make the installation as good as possible. View local regulations as the *minimum acceptable standard*, not the maximum; go one better. In other words, build the antenna installation like a brick outhouse.

Both of these mechanical integrity issues become extremely important if a problem develops. For example, suppose a wind or snow storm wrecks the antenna, plus a part of your house. The insurance company will not pay out (in most cases) if your local government requires inspections and you failed to get them done. Make sure the mechanical and/or electrical inspector (as required by US law) leaves a certificate or receipt proving that the final inspection was done. It could come in handy when disputing with the insurance company over damage.

A quality installation starts with the selection of good hardware for the installation. Any radio/television parts distributor who sells television antenna hardware will have what you need. I used Radio Shack stand-off brackets, ground pin, and a 5.8 m telescoping mast. Wherever you buy, select the best-quality, strongest material that you can find. Opt for steel masts and brackets over aluminum, no matter what the salesperson behind the counter tells you. Keep in mind that, although salespeople can be knowledgeable and helpful, you, not they, are responsible for the integrity of the installation. In my own case, I found that the 5.8 m mast was considerably sturdier at 4.6 m than when fully up, so I opted to use less than the full length because the installation is unguyed. Because I have never trusted the little cotter pin method of securing the mast at the slip-up height, I drilled a single hole through both bottom and slip-up segments (which telescope together), and secured the antenna mast with a 8 mm stainless steel bolt. The bolt was 'double nutted' in order to ensure that it did not come loose over time.

The television mast is set on a ground mounting pin/plate that is set into a 76 cm deep (local frost line regulations required only 71 cm) fencepost hole filled with concrete. The top end of the mast was secured to the roof overhang of the house (see Figure 2.23). That overhang was beefed up with 5 × 20 cm kiln-dried lumber that was bolted between two, 60 cm center, roof rafters. I felt it necessary to do that because the roof is only plywood, and the gutter guard is only 2.5 × 20 cm lumber (and is old), and the soffits are aluminum. There was not enough strength to support a 12 m lever arm, whipping around in a 35 knot wind.

Wind can be a terrible force, especially when acting on the 'sail area' of the antenna through a 7.5–12 m lever arm. A shabby installation will tear apart in wind, causing the antenna to be damaged, damage to the house, and destruction of the installation. That is why I recommend 'brick outhouse' construction methods. Over the 33 years I have been in amateur radio, I have seen a lot of verticals toppled over. Except for a few shabby models that were so poorly built that they should not have been on the market in the first place, all of these failed installations were caused by either poor installation design or poor-quality materials.

Antenna erection methods

In this section we will look at some methods for erecting wire antennas. It is recommended that you also consult Chapter 3 to find out how to make connections to wire antennas and the usual fittings (end insulators, center insulators, BALUN transformers, etc.). Chapter 3 also deals with the type of wire used for antenna construction. The ideas in this chapter are not the be-all-end-all discussion, but rather points of departure of representative methods. Your own innate intelligence can figure out other methods suitable for your own situation (keeping safety foremost in your plans).

Figure 2.19 shows one of the most basic methods for installing a wire antenna. A mast is installed on the roof of the house, or on the side of the building close to the roof. Alternatively, the mast could be replaced by some

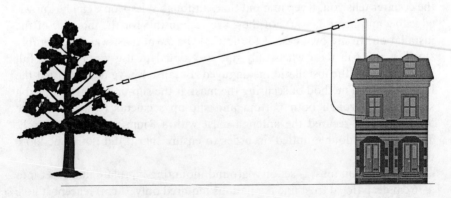

FIGURE 2.19

other support on the house. The other support is a convenient nearby tree. It could also be another building or a mast erected on the ground. The antenna is supported between two end insulators, that are in turn supported by ropes connected to the mast and the tree. For a simple random length Marconi antenna, as shown here, the direction of the wire can be up, down, or horizontal, depending on the convenience of the situation (horizontal is preferred but not always easily attained). The down-lead to the receiver or antenna tuner is run through a convenient window. There are special straps and fittings for making the connection under the window (see your radio dealer).

The installation shown in Figure 2.20 is a half-wavelength dipole, although it could represent a variety of doublet antennas. As with the previous case, the supports can be masts, a building, or a convenient tree. The installation is very much like that in Figure 2.19. The goal in making this installation is to make the antenna horizontal, with the feed-line coming away from the antenna center at a right angle for as far as is practical.

The installation in Figure 2.21 is for a dipole, but it must be a small dipole (either inductively loaded or for a high frequency), or built over a very wide roof. Both end supports are on the roof, or on the walls at either end of the roof. The coaxial cable comes away at a right angle, and then passes through a window. A vertically polarized dipole is shown in Figure 2.22.

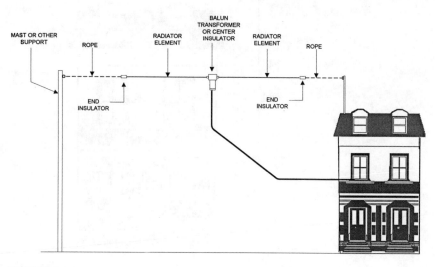

FIGURE 2.20

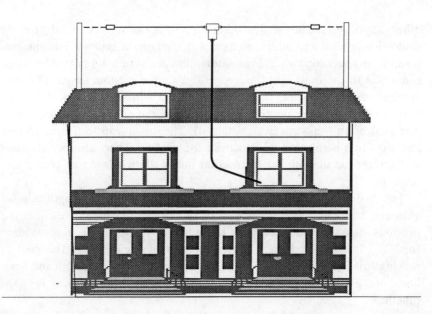

FIGURE 2.21

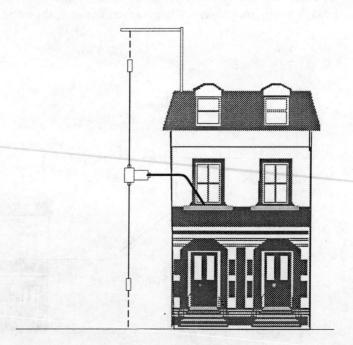

FIGURE 2.22

Figures 2.23A and 2.23B show methods for installing quad loops and delta loops, respectively. As above, the end supports can be a tree, a mast, or a building. It is necessary to support both the top and bottom of loops. The weight of the BALUN transformer and coaxial cable is not sufficient to keep the antenna from whipping around in the wind, and possibly doing some damage to either itself or other structures (not to mention the odd passerby). The bottom support can be either separate ropes to the end supports (Figure

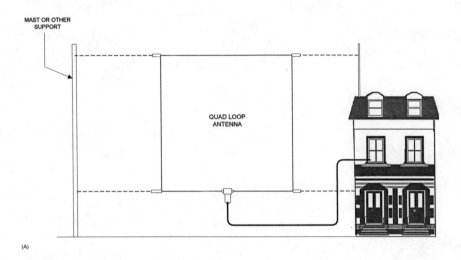

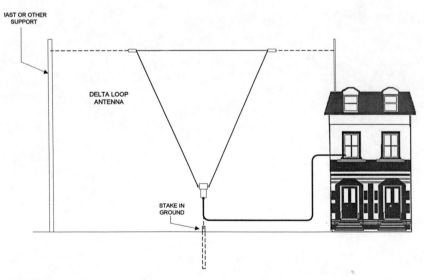

FIGURE 2.23

2.23A), or to a stake in the ground (Figure 2.23B). Make sure that the stake is secure against pulling out under the stress of wind blowing against the sail area of the antenna.

Now, let us go to Chapter 3 to talk about wire and connections.

CHAPTER 3

Wire, connection, grounds, and all that

In this chapter we will take a look at the types of wire useful for antenna elements, the different types of transmission lines, and grounding. The wire used in an antenna project can make or break the results, especially the reliability of the project over time. Antennas, especially outdoor antennas, are exposed to a lot of stress and environmental factors that tend to deteriorate both the performance and life expectancy.

TYPES OF WIRE

There are basically two types of wire used for the radiating element and other parts of antennas: **solid** (Figure 3.1A) and **stranded** (Figure 3.1B). The solid type of wire is made of a single thick strand of copper or copper alloy. Pure copper is rarely used because of both cost and strength problems, while aluminum is almost never used because it does not take solder without special, high-cost, industrial equipment. Solid wire is extruded from a single piece of copper. This is the type of wire used most often in household electrical wiring. In its smaller sizes, it is also the type of wire most commonly used as hook-up wire in electronic construction projects.

The stranded form of wire is made of two or more (sometimes many more) lengths of solid wire twisted together (Figure 3.1B). The number of strands used to make the wire sets the flexibility of the wire. Coarse stranded wire has fewer strands, and is less flexible than wire of the same overall diameter that is made up of many more strands of finer wire.

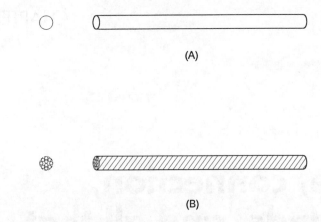

FIGURE 3.1

Antennas should normally be built with stranded wire. The reason is that the stranded wire stands up better to the constant flexing an antenna experiences, especially in windy areas. All wire will begin to stretch because of both gravity and wind. Solid wire will stretch rather rapidly, and eventually come to a point where it is too thin at the strain point to hold its own weight – and will come tumbling down. Stranded wire, on the other hand, withstands the normal stresses and strains of antenna installations much better, and will stay up a lot longer. Stranded wire antennas may come tumbling down as well, but in general it takes a lot more time and effort for the stranded wire antenna to break, compared to the solid wire type.

WIRE SIZES

Both stranded and solid wire comes in certain standard sizes. The specifications for each size wire were set basically for their ability to carry certain standard loads of direct current or the 50/60 Hz alternating current (AC) used in power systems. Table 3.1 gives some of the standard wire sizes in the range normally used by antenna constructors. In most cases, it is a good idea to stick to sizes around AWG 14 (SWG 16) for antenna work. The larger sizes are expensive, and difficult to work (most antenna hardware assumes AWG 14 or SWG 16, unless otherwise marked). The smaller sizes, on the other hand, are nice to use but often provide poor reliability – they break easily. The trade off is to use wires, of whatever standard size, that are in the 1–2.2 mm diameter region.

TABLE 3.1

AWG (USA)	Diameter		Nearest SWG (UK)
	mil[a]	mm	
8	128.5	3.264	10
10	101.9	2.588	12
12	80.8	2.053	14
14	64.1	1.628	16
16	50.8	1.291	18
18	40.3	1.024	19
20	32.0	0.812	21
22	25.3	0.644	24

[a] $1 \text{ mil} = \frac{1}{1000} \text{ inch}$.

COPPER-CLAD WIRE

Pure copper wire is not terribly good for radio antenna work because it is too soft to withstand the rigors of being outdoors, or supporting its own weight plus the weight of the fixtures and transmission lines used. As a result, one needs to find a better wire. Some people will use copper wires of harder alloys, but even that type of wire is not all that good. Others buy 'hard drawn' copper wire, and while it is better than straight electrical wiring copper wire it is not the best suited to antenna construction. The best wire for radio antennas is **copper-clad steel wire**, perhaps the most famous brand of which is Copperweld. This type of wire is made of a solid steel core coated with a copper layer. It looks like ordinary copper wire on the outside. The steel provides strength and thereby increases the reliability of the antenna installation. But steel is not a terribly good electrical conductor, so the manufacturer places a layer of copper over the outside of the wire.

The standard wire sold with the better antenna kits and by the better sources is copper-clad steel, stranded wire of AWG 14 (SWG 16) size. Each strand is copper-clad steel, and the strands are wound together to form the antenna wire.

The copper-clad wire is better for the purposes of strength, but there is a workability problem with it: copper-clad steel-core wire knots easily. If you get a knot or kink (even a minor bend) in it, that feature is always present from then on. Even if you attempt to straighten it out, it will not come anywhere near the original state. Although ordinary copper wire also possesses this attribute, it is not nearly as severe as with copper-clad steel-core wire, and can often be worked into a negligible size when it does occur. The trick is to make sure that the wire does not knot or kink during the installation process (and that can be a monumental task!).

One fair question regarding copper-clad steel-core wire concerns its resistivity. After all, any basic electrical text tells us that the cross-sectional area of the wire sets the current-carrying capacity. That is true for direct current (DC), and power line frequency AC, but only partially true for higher-frequency AC. Radio signals cause AC currents to be set up in the antenna wire, and because their frequency is so much higher than power line AC the currents flow only near the surface. The operating principle here is called the **skin effect**.

SKIN EFFECT

The skin effect refers to the fact that AC tends to flow on the surface of a conductor. While DC flows in the entire cross-section of the conductor, AC flows in a narrow band near the surface (Figure 3.2). Current density falls off exponentially from the surface of the conductor toward the center (inset to Figure 3.2). At the **critical depth** (δ), also called the **depth of penetration**, the current density is $1/\varepsilon$, or $1/2.718 = 0.368$, of the surface current density. The value of δ is a function of operating frequency, the permeability (μ) of the conductor, and the conductivity (s).

WIRE SIZE AND ANTENNA LENGTH

The size of the wire used to make an antenna affects its performance. For amateur radio operators, one concern is the current-carrying capacity. For normal amateur radio RF power levels, the AWG 14 wire size works well,

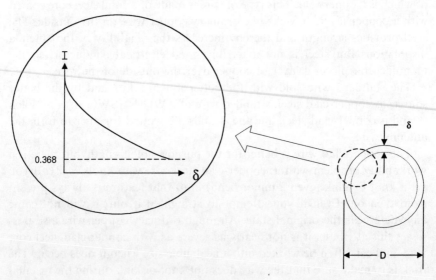

FIGURE 3.2

although those tempted to use smaller diameters with the higher legally permitted powers are forewarned: *it is not such a good idea*.

Both short-wave listeners and amateur radio operators need to be aware of another problem with the wire: the **length to diameter ratio**, or as it is sometimes specified, the **signal-wavelength-to-wire-diameter ratio**, determines the **velocity factor** (VF) of the conductor. The VF is defined as the velocity of the wave in the medium (i.e. wire) compared with the velocity in free space (which is c, the speed of light).

Figure 3.3 shows the velocity factor as a function of the L/D ratio. Wire antennas tend to have very high L/D ratios, especially at high frequencies. For example, a half wavelength at 10 MHz is 15 m (or 15 000 mm), while AWG 14 wire is about 1.63 mm diameter, resulting in $L/D = 15\,000/1.63 = 9200$. The VF of this L/D ratio is around 0.97–0.98. At very high/ultrahigh frequencies (VHF/UHF), on the other hand, the L/D ratio declines. At 150 MHz, the AWG 14 wire half-wavelength antenna has a physical length of 1500 m, making $L/D = 1500/1.63 = 920$. The VF drops to about 0.95–0.96. At 450 MHz, the L/D ratio drops to 200, and the VF runs to 0.94 or so.

So what? Why is the VF important in antenna design? Most of the antennas in this book are designed to be resonant, which means that the element lengths are related to wavelength: quarter wavelength ($\lambda/4$), half

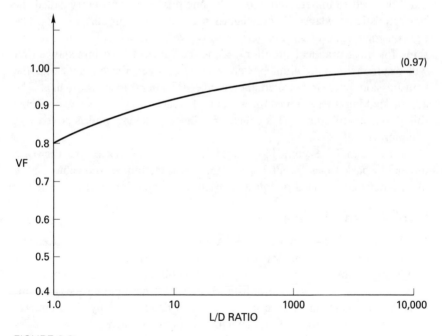

FIGURE 3.3

wavelength ($\lambda/2$), full wavelength (λ), and so forth. These lengths are required electrical lengths, and are only found in free space with perfect connectors. In antennas and transmission lines, the VF shortens the physical length required to achieve a specified electrical length. To find the physical length, multiply the required electrical length by the velocity factor. For example, a half wavelength in free space is defined as $150/F_{MHz}$, while in a wire it is $(150 \times VF)/F_{MHz}$. The result is that antenna elements must be shortened a small amount compared to the free space length. If you cut some antennas to a physical length that is comparable to the wavelength of the desired resonant frequency, then you will find that the actual resonant frequency is a tad low – meaning that the antenna is a little bit too long. This effect is almost negligible on the 80 m band, but becomes more substantial at 10 m and downright annoying at 2 m.

TRANSMISSION LINES

The wire between the antenna and either a transmitter or receiver is called the **transmission line**. Although it is tempting to think of the transmission line as a mere wire, it is actually equivalent to a complex inductor–capacitor network. The details of the transmission line network are a bit beyond the scope of this book, but can be found in antenna engineering textbooks for those who are interested. The main aspect of the transmission line network that you need to understand is that the line possesses a property called the **characteristic impedance** or **surge impedance**, which is symbolized by Z_0. The rigorous definition of Z_0 can get a bit involved, but it breaks down to one thing for most practical situations: Z_0 is the value of load impedance that will result in the maximum power transfer between the antenna and the transmission line, or between the receiver/rig and transmission line. The act of making these impedances equal to each other (a very desirable thing, as you will find out) is called **impedance matching**, and is covered in Chapter 11.

Transmission lines come in a bewildering variety of species, but there are only a few basic types: **parallel open lead**, **twin-lead**, and **coaxial cable** are the main types that we need to consider. Figure 3.4 shows all of these varieties.

Parallel open-wire line

Parallel line is shown in Figure 3.4A. It consists of a pair of wires, run parallel to each other, and separated by a constant space. Insulating spacers (made of ceramic, nylon, or some other material) are used in practical parallel open line to keep the distance between the conductors constant. The characteristic impedance of open line is determined by the diameter (d) of the conductors and the spacing (s) between them. Typical values run from 300 to 1000 ohms. Although you can calculate the spacings needed

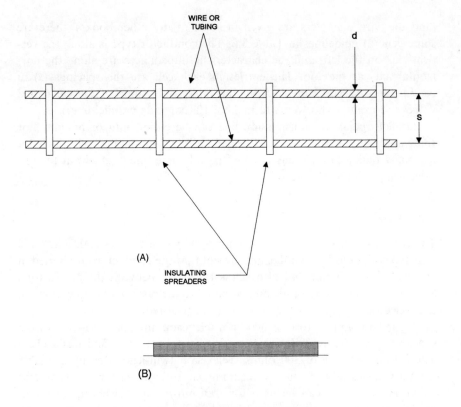

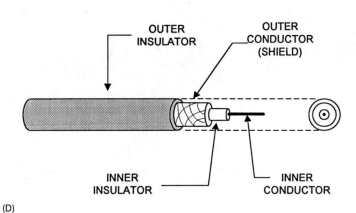

FIGURE 3.4

WIRE, CONNECTION, GROUNDS, AND ALL THAT **55**

(and the wire diameters are given in some of my other books), there are some general guidelines in Table 3.2. The conductor type is along the vertical axis on the left, and the characteristic impedances are along the horizontal axis at the top. The entries in each cell are the spacings (s) in centimeters. As you can see from the table, only some of the calculated values are useful (who wants to make a 129 cm wide parallel line?).

Parallel open-wire transmission line can be either built or bought. For some applications, such as tuned feeders, it might be the best selection. But for many balanced antennas that use parallel line one might want to consider using twin lead instead.

Twin-lead line

The twin-lead form of transmission line is shown in Figures 3.4B and 3.4C. This type of line is like parallel line, except that the conductors are buried in an insulating material that also spans the space between the conductors. Spacings are shortened a little bit compared to the open-wire variety because the dielectric constant of the insulating material reduces the VF.

The most common form is 300 ohm television antenna twin-lead, of the sort shown in Figure 3.4B. This line is about 1 cm wide, and has a characteristic impedance of 300 ohms. Television twin-lead is easily available from a wide variety of sources, including shops catering for television and video customers. It can be used for transmitting at powers up to about 150 W, but even that level is questionable if the line is of a cheaper variety. Receive-only installations have no such limitation.

Television 300 ohm twin-lead may be solid as shown, or may have oblong or rectangular holes cut into the insulation separating the conductors. This tactic is used to reduce the losses inherent in the transmission line. In other cases, the center insulating material will be rounded and hollow. This

TABLE 3.2

Conductor	Characteristic impedance, Z_0 (ohms)				
	300	400	500	600	800
8 AWG	2.0	4.6	10.6	24.4	129
10 AWG	1.6	3.6	8.4	19.3	102
12 AWG	1.25	2.9	6.7	15.3	81
14 AWG	1.0	2.3	5.3	12.2	64
16 AWG	0.80	1.8	4.2	9.6	51
18 AWG	0.63	1.4	3.3	7.6	40
20 AWG	0.50	1.1	2.6	6.1	32
22 AWG	0.40	0.90	2.1	4.8	25

approach to design also reduces the losses, and is especially effective at UHF.

The type of twin-lead shown in Figure 3.4C is 450 ohm line. This type of line is specialist, and must be sought at shops catering for amateur and commercial radio operators. It is about twice as wide as the 300 ohm variety, and almost always comes with rectangular or oblong loss-reducing holes cut into the insulation. The 450 ohm twin-lead is considerably heavier than 300 ohm line, and uses larger conductors. As a result, it can handle higher power levels.

Twin-lead is wonderful stuff, but can cause problems in some installations. The problems are especially severe at VHF/UHF, but nonetheless exist at HF as well. For example, care must be given to where the line is run: it should not be in close proximity to metal structures such as aluminum house siding, rain gutter downspouts, and so forth. It is also susceptible to picking up local electrical fields, which means a possibility of interference problems. One solution is to use coaxial cable transmission line.

Coaxial cable

This type of cable is shown in Figure 3.4D. The name 'coaxial' comes from the fact that the two conductors are cylindrical (the inner one being a wire) and share the same axis. The inner conductor and outer conductor are separated by the inner insulator. The outer conductor is usually a braided wire or foil shield, and is covered with an outer insulating material.

The characteristic impedances of commercial coaxial cables are of the order of 36–120 ohms, with 52 and 75 ohms being the most common. The specific impedance value is a function of the conductor sizes and the spacing between them. There are several standard types of coaxial cable, and these are available in many subtypes (Table 3.3) that use similar numbering.

The terms 'thick' and 'thin' would have been replaced with actual dimensions if this book were being written only a few years ago, but because of the wide array of different types now available it was decided to use these designations instead. One reason is that there are some RG-8/U specialist cables used by amateur radio and commercial antenna installers that are smaller than regular RG-8/U but larger than RG-58/U or RG-59/U.

TABLE 3.3

Type	Characteristic impedance, Z_0 (ohms)	Size
RG-8/U	52	Thick
RG-11/U	75	Thick
RG-58/U	52	Thin
RG-59/U	75	Thin

One of the differences between the different forms of coaxial cable is the insulating material used for the center insulator. Several types are used: **polyethylene**, **polyfoam** and **Teflon**. A principal effect of these materials is to change the VF of the line.

Velocity factor

The VF (denoted by V in some textbooks) of transmission line is calculated in exactly the same way as the VF of other conductors: it is the ratio of the velocity of the signal in the transmission line over the velocity of the signal in free space. The free space velocity is the speed of light (c), or about 300 000 000 m/s. The VFs of several popular transmission lines are shown in Table 3.4.

The effect of the VF is to reduce the physical length of the coaxial cable that is cut for a particular frequency. And this foreshortening can be quite drastic. For example, if polyethylene cable is cut for an electrical half wavelength at, say, 5 MHz, the actual length $L = (150 \times 0.66)/5 = 19.8$ m, while the wavelength of the signal in free space is 1.52 times longer, or 30 m.

CONNECTING THE TRANSMISSION LINE TO THE RIG OR RECEIVER

The method used to connect the transmission line to a receiver or transmitter depends on the type of output/input connector that is provided for the antenna, and the type of transmission line. In the discussion to follow, the illustrations show a receiver, but apply equally well to transmitters. The only difference is that parallel output on transmitters tends to be large ceramic feedthrough insulators, while on receivers, parallel line inputs tend to be small screw terminals on a plastic, Bakelite or other insulating carrier.

Figure 3.5 shows the basic coaxial connector on a receiver. Although an SO-239 'UHF' connector is shown, and may well be the standard, you may also find BNC, Type-N, or RCA 'phono' plug connectors used on some rigs (I've even seen phono plug connectors used on 100 W transmitters, but that

TABLE 3.4

Type of line	VF
Parallel open-wire	0.95–0.99
300 ohm twin lead	0.82
300 ohm twin lead (with holes in insulation)	0.87
450 ohm twin lead	0.87
Coaxial cable	0.66–0.80
Polyethylene	0.66
Polyfoam	0.80
Teflon	0.72

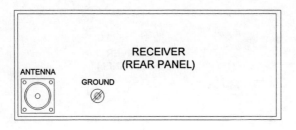

FIGURE 3.5

is not the recommended practice!). In this type of installation, the coaxial cable from the antenna (or any antenna tuning unit that might be present) is fitted with a connector that matches the chassis connector on the receiver. The threads of the male PL-259 connector are tightened onto the female SO-239 connector. Or, in the case of BNC connectors, they are pressed on and then rotated a quarter of a turn (it is a bayonet fitting), and phono plugs are just pushed into the socket.

If the receiver or transmitter is fitted with a ground connection, then it should be used. In short order we will discuss what a 'good ground' means in this context, but for the time being rest assured that you will need a heavy wire to the 'good ground' (whatever it is). If your receiver or transmitter lacks a ground terminal, then wonder why and consider that next time you select a new model.

Although most modern receivers are fitted with coaxial connectors for the antennas, there are a few models (including many that are old but still useful) which are fitted with a **balanced input** scheme such as that shown in Figure 3.6. In the case of Figure 3.6A, there are three screw terminals: A1, A2, and G (or the variants 'ANT1', 'ANT2', and 'GND'). The A1 and A2 terminals form a balanced input for a parallel transmission line, while G is the chassis ground. When an unbalanced antenna transmission line, such as a single-wire downlead (Figure 3.6B) is used, then the line goes to either A1 or A2, while the other is connected to G (in the case shown, A1 receives the antenna downlead, while A2 and G are connected together and then connected to the earth ground). Some models are equipped with only two terminals (Figure 3.6C), and in that case the selection is easy: connect the downlead to A1 (or just 'A' or 'ANT'), and the ground lead to 'G' or 'GND.' When parallel line, twin lead, or any other form of balanced line is used (Figure 3.6D), then connect 'G' or 'GND' to ground, and the two conductors of the transmission line to A1 (or ANT1) and A2 (or ANT2).

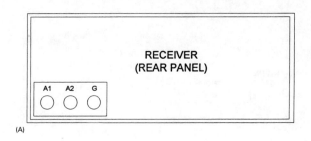

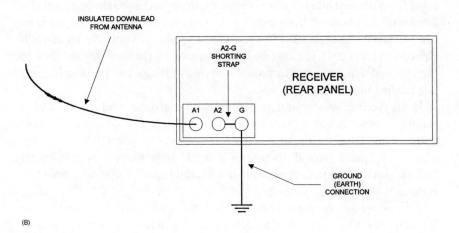

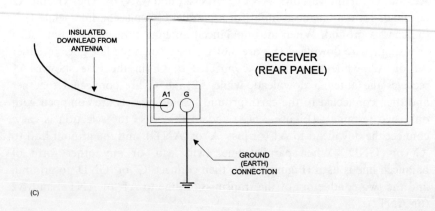

FIGURE 3.6

60 ANTENNA TOOLKIT

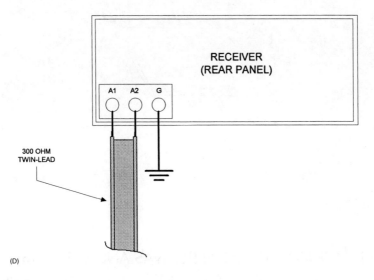

FIGURE 3.6 (continued)

CONNECTING WIRES TOGETHER

Wire antennas often require that two or more wires be spliced together (for example, two antenna elements or an antenna element and a downlead). There are right ways and wrong ways to accomplish this task. Let us look at some of the right ways.

All of the correct methods (only a few of which are shown here) have two things in common: they are (a) electrically sound, and (b) mechanically sound. Electrical soundness is created by making a good electrical connection, with the wires bound tightly together, and protected from the elements. The electrical soundness of the joint is made better and longer lasting by soldering the connection.

Figure 3.7 shows how to make the connection mechanically sound. Overlap the two wires being spliced by a few centimeters (Figure 3.7A), and then wrap each one onto the other (Figure 3.7B), forming seven or more tight turns (Figure 3.7C). Once the splice is made, then solder both knots to form a better electrical connection.

One mistake made by novice antenna builders is to assume that the purpose of soldering is to provide mechanical strength. This is false. Solder only improves and keeps good the electrical connection. Solder for radio work is a 50/50 or 60/40 lead/tin mixture with a resin core (NEVER acid core!). It is soft and has no inherent mechanical strength. Always depend on the splice to provide mechanical strength, and NEVER use solder for strength.

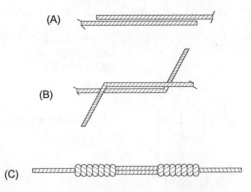

FIGURE 3.7

CONNECTING WIRES TO ANTENNAS AND SUPPORTS

The methods for connecting wires and antenna elements are basically the same as above, but with differences to account for the types of hardware being used in the antenna construction. Figure 3.8 shows how the ends of the wire antenna are supported. **An end insulator** is used. These insulators may be of glass, ceramic, or a synthetic material (nylon is common), but all will have a body and two holes. The wire from the antenna element is passed through one hole, and then looped back onto itself to form seven or more turns. As in the previous case, the splice may be plated with solder to prevent corrosion from interfering with the electrical integrity of the connection. The other end of the insulator is fitted to a rope that is run to a tree, edge of the building, or a mast of some sort. The rope should be treated similarly to the wire in that it should be wrapped over on itself several times before being knotted.

If you use an antenna that has a single-wire downlead, then you should connect it to the antenna in the manner of Figure 3.9. The downlead is almost always insulated wire, even when the antenna is not insulated. The insulation of the downlead prevents it being shorted out somewhere along its run. The scheme in Figure 3.9A shows how most people connect the downlead (and, indeed, how I have connected them quite often). This is not the best practice, although it will work most of the time. The problem is that the connection is stressed by the downlead, and may break. More

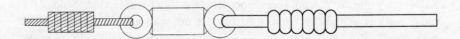

FIGURE 3.8

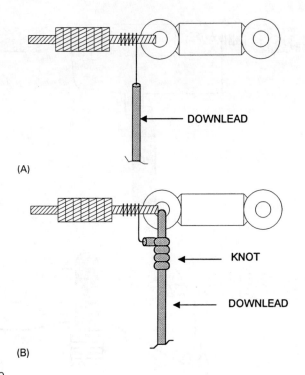

FIGURE 3.9

than once I have been off-the-air due to a spliced downlead breaking right at the junction with the antenna wire. A better solution is shown in Figure 3.9B. In this installation, the downlead wire is passed through the hole in the insulator, and then wrapped back on itself three or four times before being spliced to the antenna wire in the usual manner. This arrangement provides some strain relief to the downlead, and that (usually) translates into a longer life expectancy.

Some antennas, such as the center-fed half-wavelength dipole (a very popular wire antenna), use a center insulator to separate the two halves of the antenna. Coaxial cable is connected such that the inner conductor is spliced to one antenna element, and the shield (outer conductor) is connected to the other antenna element. Good practice is to splice the shield to the element closest to the house, if this is a factor. The method shown in Figure 3.10A will work, but it is also a bit foolish (I know, I know, I have used it and even recommended it in the past). The problem is that the center conductor is quite weak, and may be even smaller than single-wire downleads. As a result, it has no strength. To make matters worse, coaxial cable weighs more length-for-length than single-conductor downleads (or even some twin lead). The connection of Figure 3.10A will work, but only for

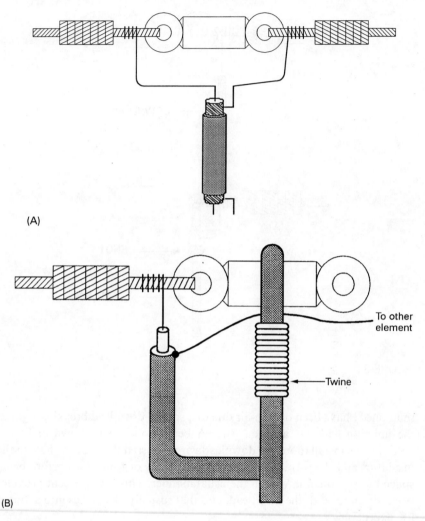

FIGURE 3.10

a while – it will surely come down prematurely. A better scheme is shown in Figure 3.10B. In this case, the coaxial cable is wrapped once around the center insulator, and secured with a length of twine or string. The electrical connections are made in exactly the same manner as shown in Figure 3.10A.

Perhaps the best solution is to use a manufactured center insulator (Figure 3.11). These devices come in a wide variety of sizes and shapes, so this illustration represents a lot of different devices. What they all share in common, however, are means for strain relieving the antenna element wires and coaxial cable. Some devices may be straight insulators, while others may

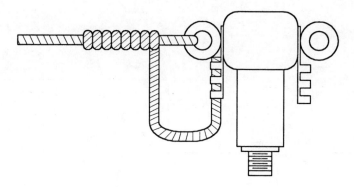

FIGURE 3.11

contain BALUN transformers inside (of which, more shortly). Some models include special strain relief features for the coaxial cable, or they may have a screw-eye on top to secure the antenna center to a mast or other support.

The electrical connections between the center insulator and the wire antenna elements are done in the usual manner (see above), but the ends of the wires are usually attached to eyelet connectors, which are crimped and then soldered. It has been my experience that these connectors are often corroded (as seen by the dull appearance) when they arrive, so brighten them with sandpaper or steel-wool before attempting to make the connections permanent.

LIGHTNING ARRESTORS

Lightning is messy stuff, and can create havoc with your radio – and might set your house on fire. But there are some things you can do to protect yourself. But first, let us take on one myth right away: contrary to popular belief, *antennas do NOT inherently attract lightning*. Lightning will usually strike the highest object around, and if that happens to be an antenna, then the antenna will take a hit. And that hit could set the house on fire, and the radio will be damaged or even destroyed. The lightning would have come anyway, and if it takes out the antenna rather than your house, then it acted like a lightning rod and protected your property.

If lightning actually strikes your antenna, then it will probably wipe out your transmitter or receiver no matter what you do. But a direct hit is not needed for fatal damage to occur: even lightning overhead or striking nearby will induce a large enough voltage into your antenna to wreak havoc inside the equipment. And it may be silent damage because it is not so spectacular as the problems seen with direct hits.

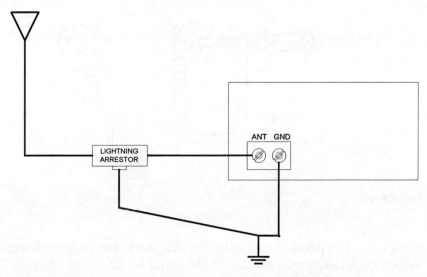

FIGURE 3.12

The key to protecting your equipment is to use a **lightning arrestor** (Figure 3.12). These devices come in several types, each suitable for coaxial line, parallel line, or single-wire downleads. Each will have at least one connection for the antenna wire, and a ground connection. It is placed in the line between the antenna lead and the receiver or transmitter, and should be mounted outside the house. There is very little that can be done for a direct hit, but I have heard tales of near hits only causing a small amount of rig or receiver damage because an effective lightning arrestor was in place. Also, keep in mind that some local laws or insurance regulations make the lightning arrestor mandatory. If you fail to use one, you may be liable for a fine from the local government, or find your homeowner's insurance not valid after a lightning episode. Lightning arrestors are cheap, and so make very reasonably priced insurance.

BALUN TRANSFORMERS

Both impedance transformation and conversion between unbalanced and balanced lines can be accomplished with a BALUN transformer (the name comes from **BALanced–UNbalanced**). Two sorts of transformers are used (Figure 3.13). The 1:1 BALUN transformer converts the balanced load to an unbalanced form (so it can be fed with coaxial cable). The impedance of the load must be the same as the source. The version shown in Figure 3.13B is a 4:1 BALUN. This transformer is used to provide not only a balanced-to-

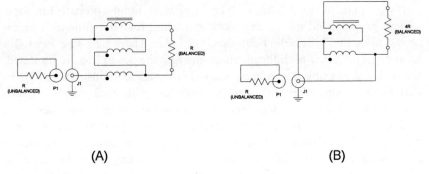

(A)　　　　　　　　　　　　　(B)

FIGURE 3.13

unbalanced conversion, but also a 4:1 impedance transformation. For example, a half-wavelength folded dipole antenna is balanced with respect to ground, and has a feedpoint impedance near 300 ohms. A 4:1 BALUN transformer converts the 300 ohm load to 75 ohms, and makes it appear an unbalanced rather than balanced load.

BALUN transformers can be easily built using toroidal coil forms or ferrite rods. The 1:1 BALUN is wound in the trifilar style, i.e. the three wires are wound side by side so that all three coils interleave with one another. The 4:1 BALUN is wound in the bifilar manner. If you do not wish to make your own BALUN transformers, then you will find that there are many commercially available BALUN transformers on the market.

VHF/UHF scanner users will find few BALUNs specially for their needs, but there are plenty of devices available. The television antenna transformers are not usually called 'BALUNs,' but when you find a transformer that converts 75 ohm coaxial cable to 300 ohm twin-lead, then it is (rest assured) a BALUN, and will work over the entire VHF/UHF range unless marked otherwise. BALUN transformers are bidirectional devices, so can be used either way around.

SAFETY

Antenna safety is an absolute must. First, let me make one thing absolutely, positively, clear in your mind:

DO NOT EVER TOSS AN ANTENNA WIRE OVER THE AC POWER LINES!

If you plan to do this trick, then have your family prepare for some solemn praying, sad music, and slow marching, for your funeral is imminent. This action kills radio hobbyists every year. There is no safe way to do it, so do not. Even if both the antenna wire and the power line are insulated, the forces of erecting the antenna will slice through both insulations! When that happens, the wires will come into contact with each other and with you – with disastrous consequences: human bodies make lousy fuses.

The next matter is like unto the first: do not erect your antenna where it can fall onto the power lines, or other people's property, either during erection or as a result of a structural failure. Falling onto the electrical lines can make the antenna deadly, and falling onto your neighbor's property is just plain rude (and might get you sued).

Finally, do not install an antenna without assistance. I know they are lightweight, and all that, but they have a remarkable 'sail area,' and even very weak winds will increase their apparent weight. I've suffered injury myself (wrenched back) by working alone, and a friend of mine broke a pelvis and a leg falling off a roof while trying to install a television antenna.

CHAPTER 4

Marconi and other unbalanced antennas

Marconi antennas are those that are *fed unbalanced with respect to ground* (Figure 4.1). In the case of a transmitter, one side of the output network is grounded, while the other side is connected to the radiator element of the antenna. The length of the radiator element varies with type, and determines many of the operating properties of the antenna. The direction of the radia-

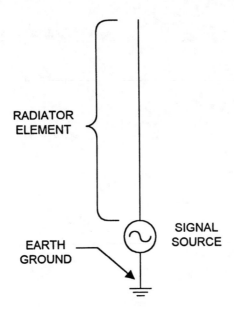

FIGURE 4.1

tor element also varies. It may be vertical as shown (indeed, vertical antennas are a species of Marconi antenna), or at a slant. In some cases, the Marconi antenna is connected directly to the rig or receiver, but in other cases an antenna tuning network is needed.

BASIC RECEIVING 'TEE' ANTENNA

Although most of the antennas in this chapter can be used on either receive or transmit, this one (Figure 4.2) is recommended for receive use only unless an antenna tuner is provided. This antenna is not particularly well thought-of, but is the antenna of choice for many first-time receiver owners, or those who live in a difficult antenna erection venue.

The radiator element is a 10–40 m long (more or less) piece of wire, preferably in the 12 AWG ($\approx$14 SWG), 14 AWG ($\approx$16 SWG), or 16 AWG ($\approx$19 SWG) size. Smaller sizes can be used, but they are not recommended for structural reasons (they break and fall down too easily).

THE HALF-LAMBDA-'TEE' ANTENNA

The antenna in Figure 4.3 is called the half-lambda-'tee' because it is one-half wavelength long (i.e. $\lambda/2$) and is a 'tee' antenna. The 'tee' is seen by the fact that a downlead section is provided, and it comes away from the radiator element at a 90° angle. The length of this downlead section is quarter wavelength. This length differs from the top-hat tee antenna used by many

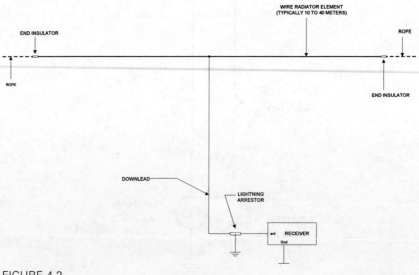

FIGURE 4.2

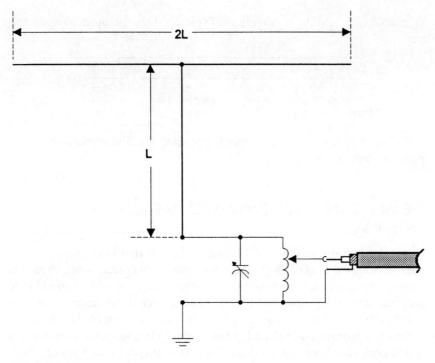

FIGURE 4.3

short-wave listeners where the radiator is half wavelength or longer and the downlead is a random length of wire. The lengths of these elements are

$$2L = \frac{143}{F_{MHz}} \text{ meters}$$

and

$$L = \frac{71.5}{F_{MHz}} \text{ meters}$$

These lengths make the antenna resonant. Examples of the proper lengths for various bands are:

Band	L	2L
60 m (5 MHz)	14.30 m	28.60 m
31 m (9.75 MHz)	7.33 m	14.66 m
20 m (14.2 MHz)	5.03 m	10.06 m

An antenna tuner is required for the half-lambda-'tee' antenna. Both the inductor and the capacitor should have reactances of 450 ohm, i.e.

$X_L = 450$ ohms and $X_C = 450$ ohms. Typical values for these components are:

Band	C	L
60 m (5 MHz)	65 pF	16.0 µH
31 m (9.75 MHz)	33 pF	8.2 µH
20 m (14.2 MHz)	23 pF	5.6 µH

For operation over a wider band, try using variable capacitor and/or inductor elements.

THE 'QUICK 'N' DIRTY' TWIN-LEAD MARCONI ANTENNA

The antenna shown in Figure 4.4 is one of the first that I ever used (when a mentor, MacIvor Parker, W4II, gave me a roll of twin-lead wire). I ran this antenna out of a basement window, and attached the top end to an old pine tree, and the middle supported by a 5 × 10 cm piece of construction lumber nailed rather unceremoniously to a tool shed on the back of the house.

Basically a quarter-wavelength 'round robin' Marconi, it consists of a 300 or 450 ohm twin-lead transmission line with a length (in meters) of 71.5/F_{MHz}. Thus, for the 60 m (5 MHz) band the length would be about 14.3 m long overall. The length of the coaxial cable feedline is not critical. The 52 ohm coaxial cable feedline is connected such that the center conductor goes to one conductor of the twin-lead, and the shield goes to the other.

This antenna is basically a species of Marconi antenna, so requires a good ground in order to work properly. At the minimum, one of the longer copper or copper-clad steel ground rods should be driven into the earth at

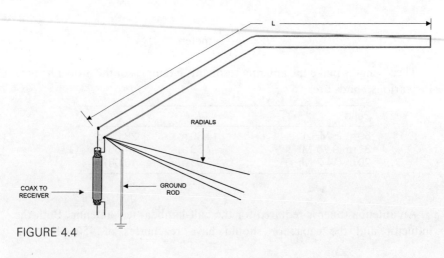

FIGURE 4.4

or very near the feedpoint. The best method is to use both a ground rod and a system of two to eight quarter-wavelength radials as a counterpoise ground.

The 'quick 'n' dirty' (or 'round robin' as my mentor called it) can be installed at an angle (as shown), although I suspect best performance occurs when the wire is straight. In the version that I had, there was only a small voltage standing wave ratio (VSWR) with one bend that put the vertical and horizontal segments at about a 120° angle.

Note the far end of the twin-lead wire (i.e. the end away from the coaxial cable). The two conductors of the twin lead are *shorted together*. I have encountered some confusion over this matter, after all a short is a short, isn't it? No, because this antenna is a quarter wavelength long, so the impedance reflected to the feedpoint by a short at the end is very high.

This antenna works better than single-conductor Marconi antennas because regular Marconi antennas suffer from ground losses. By supplying a return line, the radiation resistance is raised from 10 or 15 ohms to something of the order of 40–50 ohms – which makes it a good match to 52 ohm coaxial cable.

THE SWALLOW TAIL ANTENNA

Figure 4.5 shows two variants of the swallow tail multiband antenna. The radiator elements in both Figures 4.5A and 4.5B are cut to specific frequency bands, and are quarter wavelength ($L_{\text{meters}} = 75/F_{\text{MHz}}$). As many as needed to cover the bands of interest may be used, provided that they do not bear a 3:1 frequency ratio. The reason for the constraint is that for all antennas but the resonant one the impedances are so high that connecting them in parallel with each other does not affect the overall feedpoint impedance. However, at the third harmonic, the impedance again drops low and will load down the impedance seen by the transmission line. This is another situation where the results are more profound for amateur radio transmitters than for shortwave receivers, but it is a good idea to avoid it if possible. Besides, the antenna is actually resonant on its third harmonic, so you lose nothing.

The version in Figure 4.5A uses a pair of insulating masts, or support structures (roof of a house, tree, etc.) with a rope stretched between them. The quarter-wavelength resonant wires are spread out along the length of the rope, spaced approximately evenly.

The version shown in Figure 4.5B uses a large wooden cross-like structure. The antenna wires are connected to the cross-piece at the top end, and to the coaxial cable at the bottom end. This antenna apparently worked well for a fellow who wrote to me recently. He lives in a townhouse community that has a 'homeowners' association' of nit-picking little dictators who like

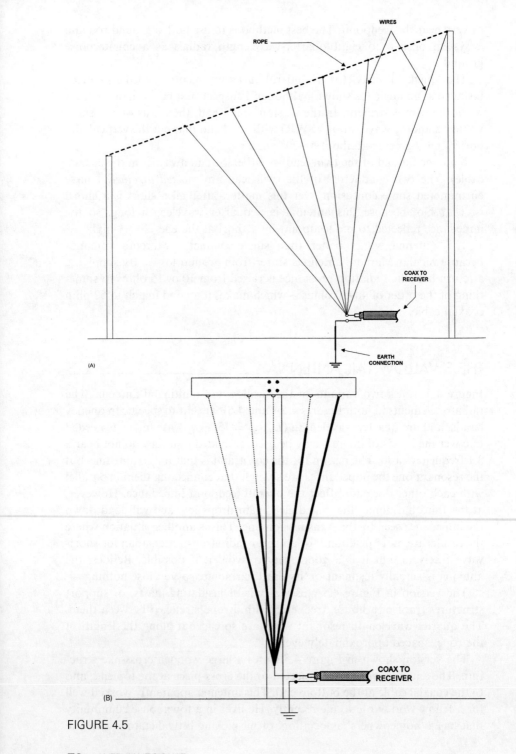

FIGURE 4.5

to tell people what they may do with their houses. One of the rules is 'no outdoor antennas' of any sort. He erected a mast in his rear garden shaped like a ship's mast and yardarm, and then convinced the busy-bodies snooping for 'the committee' (dread!) that it was somehow nautical, not radio.

RANDOM LENGTH MARCONI

There are several antennas that qualify as 'all-time favorites' with both ham operators and short-wave listeners. In the top three or four is the random-length wire antenna. This form of antenna consists of whatever length of wire that you happen to have (although there are reasonable limits!) run from the rig or receiver to a convenient elevated support (e.g. a tree, mast, or roof of a building). Figure 4.6 shows this most basic form of Marconi antenna.

Although a receiver owner can connect the random-length radiator wire to the antenna input of the receiver, the transmitter operator most certainly requires an antenna tuning unit (ATU). Even receiver operators will find this antenna a better performer when an ATU is used between the coaxial cable to the receiver and the radiator wire.

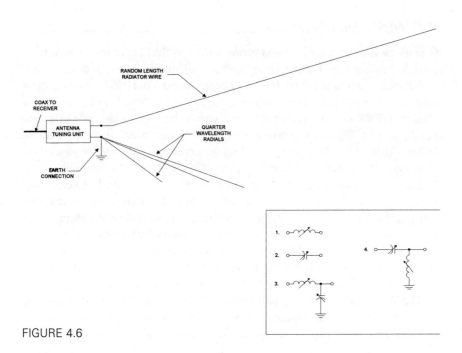

FIGURE 4.6

One requirement of this antenna is that it requires a good ground. It is essential that either an earth ground or a counterpoise ground (i.e. a system of radials) be provided. Methods of providing a good grounding system are described in Chapter 13.

There are several conditions for this form of antenna: an antenna that is shorter than a quarter wavelength, an antenna that is a quarter wavelength, and longer than a quarter wavelength. The inset in Figure 4.6 shows several options for the antenna tuning unit. At position 1 is a variable inductor. This is used when the antenna is quite short compared to a quarter wavelength. At position 2 is a variable capacitor. It is used when the antenna is longer than a quarter wavelength. When the antenna is exactly a quarter wavelength, then it may be possible to eliminate the antenna tuner. It is also possible to use the L-section couplers at positions 3 and 4. These ATUs combine the inductor and capacitor. These are used to match the feedpoint impedance to the 52 ohm coaxial cable.

Many commercial L-section couplers are designed with either switching or shorting straps that allow any of the combinations in the inset to Figure 4.6 to be realized by making some simple changes.

Receiver operators who use this antenna are well advised to use an ATU, just like transmitter operators. But there is at least one difference: you should be able to switch the ATU out of the system when you are receiving a frequency outside the limits of the tuner.

THE WINDOM ANTENNA

One of the first ham radio transmitting antennas that I ever used was not my own, but rather the antenna at my high school station in Arlington, Virginia (K4BGA, Washington-Lee High School). Although I had owned receive antennas, the K4BGA club station was one of the first I had operated. It used an HF Windom antenna running across the roof of the industrial arts building. The Windom antenna (Figure 4.7) has been around since the 1920s. Although Mr Loren Windom is credited with the design, there were actually a number of contributors. Co-workers with Windom at the University of Illinois were John Byrne, E. F. Brooke, and W. L. Everett, and they are properly co-credited. The designation of Windom as the inventor was probably due to the publication of the idea (credited to Windom) in the July 1926 issue of *QST* magazine. Additional (later) contributions were rendered by G2BI and GM3IAA (Jim MacIntosh). We will continue the tradition of crediting Loren Windom, with the understanding that others also contributed to this antenna design.

The Windom is a roughly half-wavelength antenna that will also work on even harmonics of the fundamental frequency. The basic premise is that the antenna radiation resistance varies from about 50 ohms to about 5000

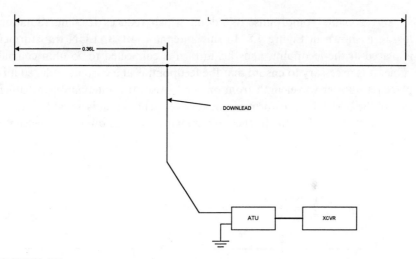

FIGURE 4.7

ohms, depending upon the selected feedpoint. When fed in the exact center, a current node, the feedpoint impedance will be 50 ohms; similarly, end-feeding the antenna finds a feedpoint impedance of about 5000 ohms. In Figure 4.7 the feedpoint is tapped away from the center at a point that is about one-third (0.36L) the way from one end, at a point where the impedance is about 600 ohms.

The feedline for the basic Windom of Figure 4.7 is an insulated length of wire. Of course, the size of the wire depends on the power level, but I suspect that No. 14 insulated stranded wire will do for most people who run less than 200 W of power (indeed, I would not like to use a Windom at high-power levels because of the 'radio frequency (RF) in the shack' problem).

The Windom antenna works well – but with some serious caveats. For example, the antenna has a tendency to put 'RF in the shack' because of the fact that it is voltage fed. This is why Windom antenna users (along with those using the random-length Marconi) get little 'nips' of RF shock when touching the transmitter chassis, or putting a lip on the microphone. Second, there is some significant radiation loss from the feedline. Finally, the antenna works poorly on odd harmonics of the fundamental frequency.

The antenna tuning unit can be either a parallel resonant, link-coupled, LC tank circuit or a π-network (i.e. high-impedance end toward the antenna).

Note that a good ground should be used with this antenna (note the ground connection at the output of the antenna tuning unit). This basically means (for most people) a 2–2.5 m ground rod, or a system of radials (see earlier discussion for random-length wire antennas).

A reasonable compromise Windom, which reduces feedline radiation losses, is shown in Figure 4.8. In this antenna a 4:1 BALUN transformer is placed at the feedpoint, and this in turn is connected to 75 ohm coaxial cable. It is necessary to ensure that the feedpoint is at a current node, so it is placed a quarter wavelength from one end. Also, the antenna on the other side of the BALUN transformer should have a length that is an odd multiple of a quarter wavelength. In the case shown it is three-quarter wavelength long.

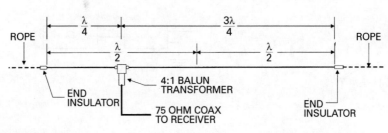

FIGURE 4.8

QUARTER WAVELENGTH VERTICAL

Perhaps the most classic form of Marconi antenna is the quarter wavelength vertical (Figure 4.9). It consists of a vertical radiator element that is quarter wavelength long, and a system of quarter wavelength radials for the ground system. The length of the radiator element is found from:

$$L_{meters} = \frac{75}{F_{MHz}}$$

where L_{meters} is the length in meters and F_{MHz} is the frequency in megahertz.

The radiator element can be made of pipe or wire. The former case, the pipe can be copper or aluminum, although I suspect that aluminum will be the choice of most readers. In the case of a wire radiator, it will be necessary to support the wire up top, so some form of mast or support is necessary. Make the mast or support out of an insulating material such as wood.

The radials are somewhat optional if you use it with a 'good ground'. But in the absence of a 'good ground', I would use at least four radials and as many as sixteen. Above sixteen the benefit of adding further radials drops off substantially.

Radials should be buried if the vertical is ground mounted (that is to prevent lawsuits over pedestrians tripping over them). In the case where the

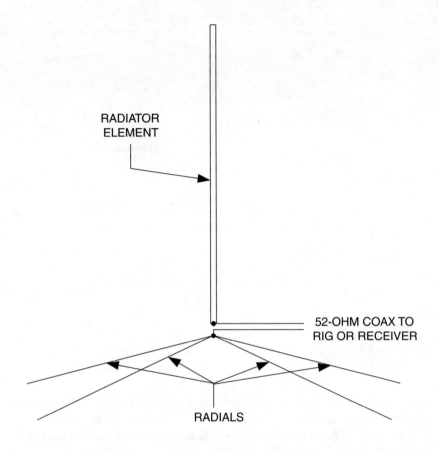

FIGURE 4.9

vertical is mounted on a mast, off ground, the radials are not optional – they are mandatory. In that case, use four to sixteen radials arranged in a circular pattern around the base of the radiator element.

Connect the system so that the radiator element is connected to the coaxial cable center conductor, and the radials are connected to the shield of the coaxial cable. Because the feedpoint impedance of the vertical antenna is 37 ohms, use of 52-ohm coax will result in a VSWR of only $50/37 = 1.35:1$. In the case where the vertical has a substantially lower impedance (as low as 2 ohms!), use a broadband transformer between the coaxial cable and the antenna.

FOLDED MARCONI 'TEE' ANTENNA

The antenna shown in Figure 4.10 is popular for receiving and transmitting on the lower bands (<7 MHz) when space for antennas is a bit limited.

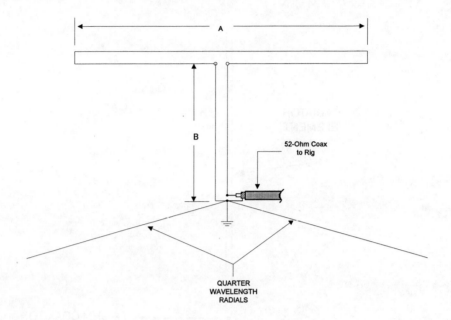

FIGURE 4.10

Although it can still take a lot of space at the lowest frequencies, it is quite a bit shorter than a half-wavelength dipole for the same frequencies.

Two pieces of 300 or 450 ohm twin-lead are used to make this antenna (A and B in Figure 4.10). The A-section is the main radiator element, and it has a length of

$$A_{meters} = \frac{82}{F_{MHz}} \text{ meters}$$

Note that section A is built like a folded dipole. At the ends of the twin-lead the two conductors are shorted together. At the center point of section A one of the two conductors is cut to accept the two conductors of the twin-lead used for section B.

Section B is vertical, and should come away from section A at a right angle. It has a length that is similar to that of A except that it is reduced by the velocity factor (VF) of the transmission line:

$$B_{meters} = \frac{82(VF)}{F_{MHz}} \text{ meters}$$

For ordinary 300 ohm twin-lead, the length of B is $0.82A$.

Note that the conductors making up *A* and *B* form a continuous loop of wire. The coaxial cable is connected such that the shield goes to one lead of *B* and the inner conductor goes to the other conductor. At the top of *B* the two conductors are connected to either side of the cut in section *A*.

As with the other Marconi antennas in this chapter, it is imperative that a good ground be used with the 'tee' Marconi. Otherwise, losses will be too great, and performance will suffer considerably.

Figure 4.11 shows the end-fed Zepp antenna. This antenna has a radiator wire that is half wavelength long at the lowest frequency of operation. It will work on harmonics of that frequency as well as the frequency itself. It will also work on other frequencies if a high VSWR can be tolerated. It is fed by 600-ohm parallel feedline and an antenna tuning unit (ATU). The ATU is used to tune the VSWR to 1:1 at the frequency of operation.

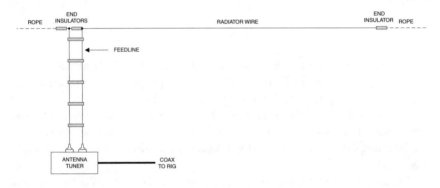

FIGURE 4.11

EWE ANTENNA

The EWE antenna emerged recently as one solution to the low-noise low-band antenna problem. Figure 4.12 shows the basic EWE antenna. It consists of two vertical sections (labeled L1) and a horizontal section (L2). The EWE looks superficially like a Beverage antenna, but it isn't. Like the Beverage it is erected about L1 = 3 meters above the Earth's surface. Unlike the Beverage, it is only L2 = 6.5 meters long at 3.5 MHz. Those dimensions make it affordable for most people.

The far end segment is terminated in an 850 ohm resistor. This resistor should be a carbon composition or metal film resistor, and never a wire-wound resistor.

FIGURE 4.12

The receiver end must be matched to the receiver's 50 Ω antenna input impedance. Transformer T1 is provided for this purpose. It has a turns ratio of 3:1 to provide the 9:1 impedance ratio required to match the 450 Ω antenna impedance to the 50 Ω receiver impedance. A powdered iron toroid core made of -2, -6 or -15 material will be sufficient. A suitable transformer can be made using a T-50-15 (red/white) core. Use about 20 turns of any size enameled wire.

The azimuthal and elevation patterns for the Koontz EWE antenna are shown in Figures 4.13, 4.14, 4.15 and 4.16. These patterns were simulated from the *Nec-WIN Basic* software available from Nittany-Scientific. The patterns in Figures 4.13 and 4.14 are based on the Sommerfield-Norton standard ground model, with the azimuth being seen in Figure 4.13 and the elevation in Figure 4.14. The same types of pattern are seen for a 'real' ground based on suburban soil and are shown in Figures 4.15 (azimuth) and 4.16 (elevation).

REVERSIBLE EWE

The EWE antenna can be made reversible by using a system such as Figure 4.17. The feedpoint and termination circuits are co-located at the receiver. Transformer T1, coil L1, resistor R1 and DPDT switch S1 are installed in a shielded metal box. The outputs of the box (i.e. center terminals of the DPDT switch) are connected to the bases of the vertical (L1) sections of the EWE antenna. According to one source, the simple resistive termination

82 ANTENNA TOOLKIT

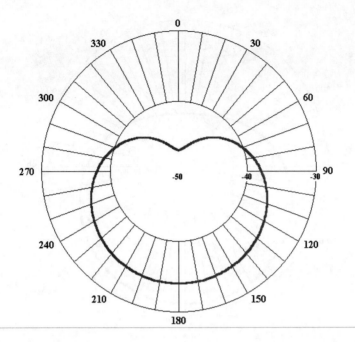

FIGURE 4.13   Low-Noise RX Antenna

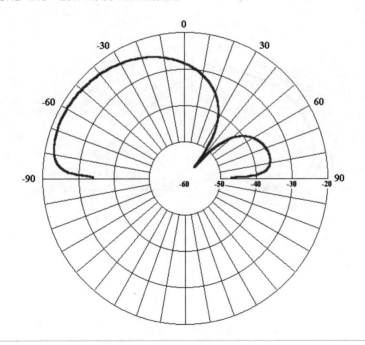

FIGURE 4.14   Low-Noise RX Antenna

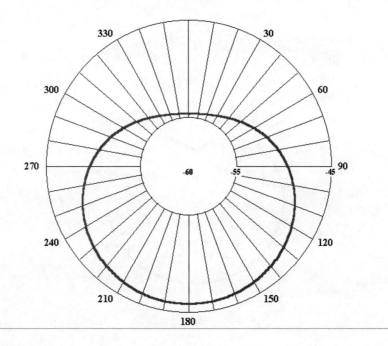

FIGURE 4.15 Low-Noise RX Antenna Rocky Soil

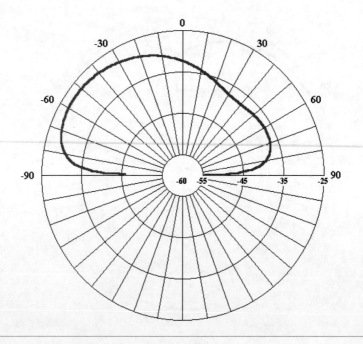

FIGURE 4.16 Low-Noise RX Antenna Rocky Soil

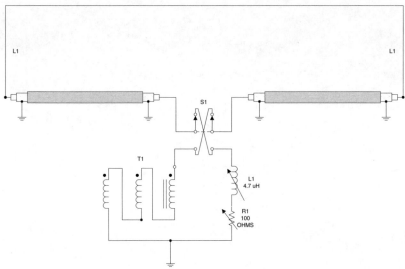

FIGURE 4.17

was not sufficient, so they added an inductive reactance in series with a resistance. This is the method used on Beverage antennas to make a steerable null, and that effect is seen on the EWE as well.

DUAL EWE ANTENNA

Figure 4.18A shows a modification of the EWE antenna that permits switchable bi-directionality. Four EWE antennas are arranged in north–south (N–S) and East–West (E–W) directions. The feedpoints (A, B, C and D) are connected to a switch circuit such as shown in Figure 4.18B. The directivity of the antenna is controlled by opening and closing the four switches (S1–S4).

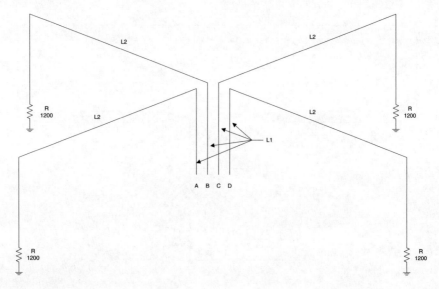

FIGURE 4.18A

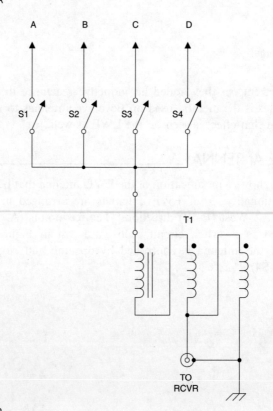

FIGURE 4.18B

CHAPTER 5

Doublets, dipoles, and other Hertzian antennas

A **doublet** antenna is any of several forms of **dipoles** (*di* – 'two') that are balanced with respect to ground. In other words, neither side of the antenna feedline is grounded. These antennas are fed in the center, which point will be either a current node or voltage node depending on the design of the particular antenna. Perhaps the most common form of dipole is the half-wavelength ($\lambda/2$) center-fed dipole, which can be fed with ordinary 75 ohm coaxial cable. It is so popular that when people use the word 'dipole' in an unqualified sense, it is the $\lambda/2$ center-fed version that is meant. Although dipoles can be made of either wire or tubing, the most common practice is to make them out of wire, especially antennas at the lower end of the high-frequency (HF) spectrum. Above about 14 MHz, however, it becomes increasingly practical to make these antennas of aluminum tubing.

DIPOLE RADIATION PATTERNS

When one uses the term 'dipole' in the common sense it is usually understood that the antenna being discussed is the half-wavelength doublet. One reason for this practice is the widespread popularity of that particular dipole. But the practice is actually erroneous because the dipole can be any size up to about two wavelengths (although some would argue that any balanced antenna, even longer than two wavelengths, is a dipole, I demur on grounds that longer antennas begin acting like the **long-wire** antenna and its relatives). Figure 5.1 shows the effect of different wavelength dipoles on the azimuthal radiation pattern of the antenna.

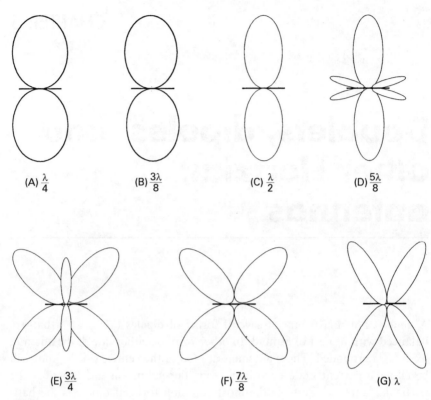

FIGURE 5.1

At $\lambda/4$, $3\lambda/8$, and $\lambda/2$ the azimuthal pattern is the classic figure '8' that is commonly associated with dipoles. The principal difference in these patterns is the particular shape (width, etc.) of the pattern, which varies with antenna size. At $5\lambda/8$ wavelength, the figure '8' persists, but becomes narrower and minor lobes begin to appear. To a receiver, these sidelobes represent responses in directions that may not be desired, and for transmitters it represents wasted energy. At $3\lambda/4$, the pattern blossoms into a four-lobe 'clover leaf' with significant minor lobes. This pattern is seen in ham radio when a half-wavelength 40 m (7 MHz) dipole is used at 15 m (21 MHz). The same physical length that is $\lambda/2$ at 7 MHz is $3\lambda/4$ at 21 Mhz. At $7\lambda/8$ wavelength the minor lobes disappear, and at 1λ the pattern thins out a bit but retains the clover leaf shape.

HALF-WAVELENGTH DIPOLE

It is often the case that a half-wavelength dipole is the first antenna that a new ham operator or short-wave listener will put up. These antennas are low cost, easy to build, and have a delightful tendency to work well with only a little muss and fuss. They are thus well suited to the newcomer.

Figure 5.2A shows the basic half-wavelength dipole. It consists of a half-wavelength radiator ('B') that is cut into two sections ('A') which are each a quarter-wavelength long. The feedpoint is the middle of the half-wavelength, and for this size dipole the feedpoint is a current node. The feedpoint impedance is therefore a minimum, and the dipole makes a good match to 75 ohm coaxial cable.

It is commonly assumed that the feedpoint impedance of the dipole is 73 ohms, but that is a nominal value which is only true at certain heights above the ground. As can be seen in Figure 5.2B, the actual impedance is a function of the height of the dipole, and at a certain height it converges to 73 ohms. At other heights it can vary from a very low impedance (near zero ohms) up to about 110 or 120 ohms.

The reason for the impedance situation is shown in Figure 5.2C. This graph shows the distribution of voltage and current along the length of the antenna element. Note that the current peaks at the feedpoint and drops to a zero at the ends. The voltage, on the other hand, peaks at the two ends and drops to a minimum at the feedpoint. One of the principal differences between different dipole designs is the issue of whether they are voltage fed or current fed.

The element lengths (in meters) of the half-wavelength dipole would normally be found from $150/F_{MHz}$ for the overall length, and $75/F_{MHz}$ for each quarter-wavelength segment. However, because of the velocity factor effects of the length–diameter ratio (which is high for wire HF antennas), and the capacitive end effects, a small foreshortening occurs, making the actual lengths closer to

$$B_{meters} = \frac{143}{F_{MHz}} \text{ meters}$$

$$A_{meters} = \frac{71.5}{F_{MHz}} \text{ meters}$$

These lengths are still only approximate, although in many cases they will be right on the money. There is nonetheless a possibility of needing to lengthen or shorten the antenna if an exact resonant frequency is needed. This is done by either noting the voltage standing wave ratio (VSWR) for the minima being at the desired resonant frequency, or by the feedpoint impedance being close to the design value.

The standard dipole is fed in the center with coaxial cable. For installations really on the cheap, ordinary lamp cord (alternating current wiring) can be used, as it looks like 75–100 ohm parallel transmission line to the antenna. However, except for receiver and the lowest power transmit

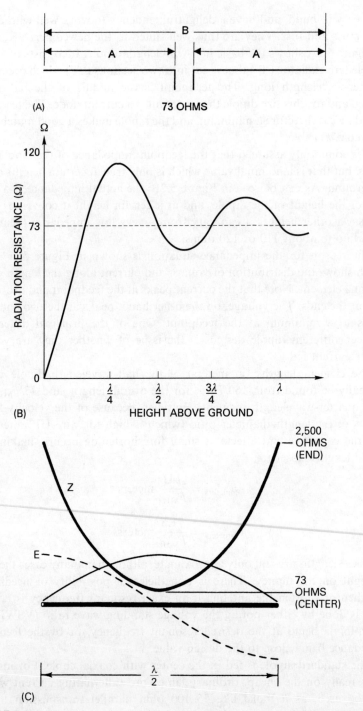

FIGURE 5.2

operations, it is definitely not recommended. Even receiver operators will fare better with coaxial cable because it has superior noise immunity.

In the usual installation, the dipole will have a center insulator to support the middle ends of the two sections, and to provide the connection spot for the transmission line. The outer ends of the two sections will be supported by end insulators and rope or heavy twine tied to a mast, roof top, or convenient tree. Antenna and radio shops sell special center insulators that provide strain relieved coaxial connections for the transmission line and solder points for the antenna elements.

To BALUN or not to BALUN, that is the question

The BALUN transformer is used to convert between balanced and unbalanced loads. In the context of the dipole, the feedpoint is balanced with respect to the ground, while the transmitter output is unbalanced with respect to the ground. What this means in plain language is that an unbalanced load has one side grounded, while a balanced load is not grounded. Some BALUN transformers also provide impedance transformation, with 4:1 being the most commonly seen. This means that a load of R across the output ports of the transformer is reflected back to the input so that 'looking into' the transformer it will seem like a load of $R/4$. This is the transformer used to reduce 300 ohms from television-style twin-lead to 75 ohms for coaxial cable.

It is common practice to use a 1:1 BALUN transformer at the feedpoint of the half-wavelength dipole antenna (Figure 5.3), even though it provides no impedance transformation (and none is needed). The BALUN transformer balances the load, causing equal but opposite phase currents to flow in the two conductors, and thereby reduces radiation from the transmission line. The effect of the feedline radiation is to distort the radiation pattern. Figure 5.4 shows the effect on the pattern, at least in simplified form. Figure 5.4A shows the 'normal' pattern, which is what we saw in Figure 5.1C. But radiation from the feedline, and unbalanced current flow, tends to cause the nulls to fill in and the main lobes to reduce a bit (Figure 5.4B). In some cases, the afflicted pattern is a lot more distorted than shown in Figure 5.4B, which is a highly simplified case. When a BALUN transformer is used, however, the resultant pattern will look a lot nearer the ideal than the distorted version, and the dipole will perform as expected.

The use of the BALUN in the half-wavelength dipole is one of those controversial things that radio enthusiasts debate seemingly endlessly, but there is no real controversy. The simple fact is that half-wavelength dipoles work better in practical situations when a 1:1 BALUN is used to interface the coaxial cable transmission line to the antenna feedpoint. The only excuse to not use a BALUN, in my not so humble opinion, is the economic one.

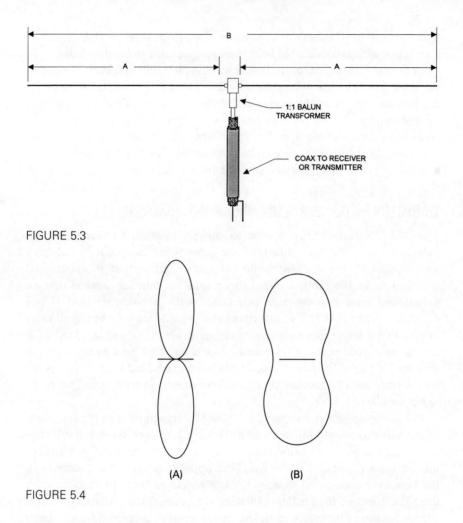

FIGURE 5.3

FIGURE 5.4

There were, after all, periods in my life when even so low a priced item as an HF BALUN was very dear indeed, so I used an end insulator at the feedpoint instead (once I even used a discarded toothbrush as a center insulator – and it even lasted a month or two before it broke).

MULTIBAND AND WIDE-BAND HALF-WAVELENGTH DIPOLES

The half-wavelength dipole is a resonant antenna, so it works best only on and near the center design frequency. The classic figure '8' pattern is found only at this frequency and certain lower frequencies (for which the antenna is $\lambda/4$ or $3\lambda/8$). The antenna also provides performance at odd integer harmonics of the design frequency, but with a multilobed pattern (see Figure 5.1 again). But there are some things that can be done to accommodate the multiband interests of typical radio hobbiests. For one thing, we could put up a different dipole for each band. But even if you have the space, the installation would look like an aerial rat's nest of wires, and it requires switching the feedlines at the rig or receiver to change bands. Not a good idea, actually.

The other approach is shown in Figure 5.5: connect dipoles for different bands to the same transmission line. One can get away with this because the feedpoint impedances of the dipoles are very high except at their resonant frequency, so do not materially affect the feedpoint impedance of whichever dipole is in use at any given time. In Figure 5.5, dipole A1–A2 works at the highest band, B1–B2 works at a middle band, and C1–C2 works at the lowest-frequency band. One precaution is to ensure that none of the dipoles is cut to a frequency that is an odd integer multiple of the resonant frequency of one of the other dipoles. If, for example, you make C1–C2 cut to 7 MHz, and A1–A2 cut for 21 MHz, then the half-wavelength 21 MHz dipole A1–A2 will be in parallel with the $3\lambda/4$ wavelength 21 MHz operation of A1–A2. The low impedances will be in conflict and the pattern will be bizarre. Note, however, that $3\lambda/4$ operation means that proper selection will yield six-band operation from this 'three-band' antenna.

A certain precaution is needed when the multiband dipole is connected to a radio transmitter. Many of the HF amateur radio bands are harmonically

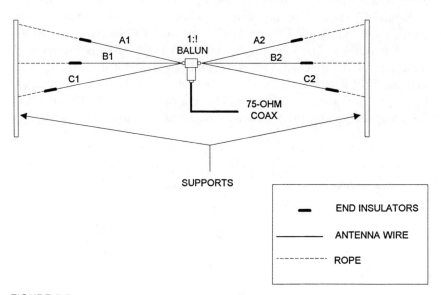

FIGURE 5.5

related. This was done, I am told by some old timers, because in the 'golden oldie' days, trapping harmonics was not as easily done as today, and amateur transmitter harmonics would fall into another amateur band, rather than onto a commercial band or public safety band. If one of the other dipoles is cut to a harmonically related band, then that harmonic looks into a resonant antenna and will radiate efficiently. If you use this type of antenna, then be certain to use a low-pass filter with a cut-off between the two bands, or use an antenna tuning unit that is not a high-pass filter (as some are), or otherwise make sure that your transmitter is harmonic free.

Figure 5.6 shows a method for producing a very wide-band dipole. This particular version was designed for use by radio science observers who listened for radio emissions from the planet Jupiter in the 16–26 MHz short-wave band. These sounds are rather strong, and can appear anytime Jupiter is above the horizon, day or night. Although a beam antenna would have been more useful in this project, the small college physics department who needed it was unable to afford a proper array, so had to make do with a simple dipole.

The idea in this design is to parallel connect three half-wavelength dipoles with overlapping frequency responses. In this particular case, the antennas were cut for 18, 21, and 24 MHz. Some amateur radio operators will use this type of antenna, but with only two dipoles, with one cut for a frequency one-third the way through the band and the other at two-thirds the way through the band. For example, in the 15 m band (21.0–21.450 MHz), the low-end antenna is cut for 21.15 MHz and the high-end antenna for 21.3 MHz.

Figure 5.7 shows how this antenna works. In Figure 5.7A we see the frequency response of a single-dipole antenna. This curve plots VSWR versus frequency. The lowest point on the VSWR curve is the resonant frequency of the antenna, and is hopefully where you designed it. Note that the VSWR climbs fairly rapidly as you get away from the resonant frequency. In Figure 5.7B we see the effect of overlapping the VSWR versus frequency curves of three antennas designed for adjacent segments of

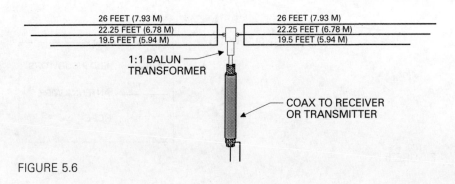

FIGURE 5.6

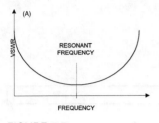 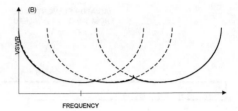

FIGURE 5.7

the band. The overall VSWR seen by the antenna is basically the lowest point on the composite curve (although there is some interaction in real antennas!).

FOLDED DIPOLES

The **folded dipole** antenna (Figure 5.8) has a wider bandwidth than the common dipole, although at the expense of a little constructional complexity. This antenna consists of a loop of wire made from a half-wavelength section of twin-lead or parallel transmission line. The most common form of folded dipole uses 300 ohm television twin-lead for the antenna element. Note that the two conductors of the twin-lead are shorted together at both ends of the radiator. The feedpoint is formed by breaking one of the two conductors at the midpoint. The feedpoint impedance of this antenna is just under 300 ohms, so it makes a good match to another piece of 300 ohm twin-lead used as a transmission line.

The overall length of the folded dipole is found from the same equation (see above) as a regular half-wavelength dipole. There may be a small amount of additional foreshortening due to the velocity factor of the twin-lead, so this length may be a tad long for the desired frequency. You can adjust to the desired frequency by trimming the same amount off of each end while monitoring the VSWR to determine the resonant frequency.

If you want to use coaxial cable to feed the folded dipole, then place a 4:1 BALUN transformer at the feedpoint. This transformer reduces the feedpoint impedance from near 300 ohms to near 75 ohms, and so makes a good match to 75 ohm coaxial cable (e.g. RG-59 or RG-11).

The folded dipole made of twin-lead can be used with moderate amounts of radio frequency (RF) power in transmitting stations, and all receiving stations, but one has to be a bit careful as power levels increase. I recall one 40 m folded dipole excited with around 300 W of RF that got uncomfortably hot to the touch after a few minutes of CW operation.

Folded dipole construction always looks easier in books than it is in real life. The problem is that the conductors used in 300 ohm twin-lead are small, and not intended to support any weight or sustain under wind forces (which

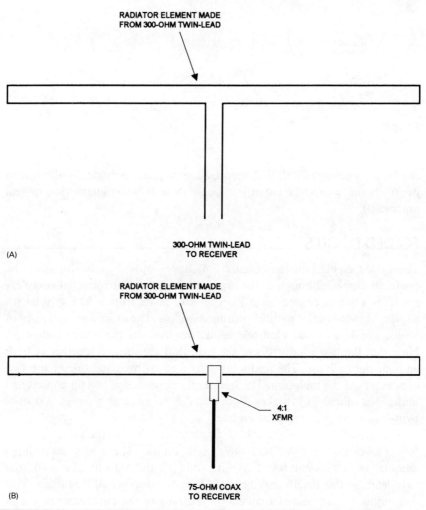

FIGURE 5.8

can be larger than most of us imagine because an antenna has a large 'sail area' in wind). As a result, folded dipoles usually break and come down either at the feedpoint or one of the two ends. The solution is to make fittings such as shown in Figure 5.9 for the center (Figure 5.9A) and ends (Figure 5.9B).

The folded dipole fittings of Figure 5.9 are made of an insulating material. I have seen similar devices made of hardwood salvaged from flooring, and then coated in polyurethene or varnish to protect against rain. Others use Plexiglass, Lexan, or other materials. I have even seen one commercial folded dipole center insulator made of nylon, but it seems to have disap-

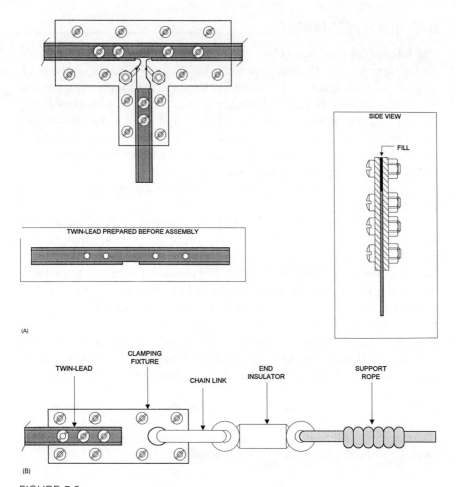

FIGURE 5.9

peared from the marketplace. In Figure 5.9A, the wire connections are soldered to solder lugs attached to brass machine screws. All other screws and nuts are made of nylon to minimize interaction with the antenna. Note in Figures 5.9A and 5.9B that holes are cut into the insulation of the twin-lead, and nylon machine screws are passed through both these holes and mating holes cut into the mounting device; matching hex nuts on the other side secure the screw (see side view inset to Figure 5.9A). This keeps the twin-lead from slipping out of the block, or at least delays the day when failure might occur.

The screw holes in the twin-lead (see inset to Figure 5.9A) can be made with an ordinary paper hole punch, although one has to squeeze a bit hard on the better grades of twin-lead.

INVERTED-'VEE' DIPOLE

The **inverted-'vee' antenna** (Figure 5.10) is a dipole that is supported in the center from a mast or other support, with the ends drooping toward the ground. As with other forms of dipole, the ends are supported using end insulators and rope. But instead of finding a convenient high spot to support the ends, the rope is tied off to a ground level support, or a stake driven into the ground if none exists. Keep in mind, however, that only the type of tent pegs that are intended for high-wind areas (i.e. the type with prongs) should be used. Otherwise, drive a long stake into the ground. The wires of this antenna have a rather large sail area, and the forces could easily pull out the ground supports over time.

The inverted-'vee' antenna is fed in the center with 75 ohm coaxial cable and a 1:1 BALUN transformer. In some cases, you will find that 52 ohm coaxial cable is a better match. The angle between the radiator elements at the top of the mast should be 90–120°. The directivity is based on the figure '8' of a dipole, but with considerable filling in of the nulls off the ends.

The length of the wire elements can be a quarter wavelength each (a half wavelength overall), or any odd integer multiple of a quarter wavelength. Because of the sloping of the elements, the actual length is of the order of 5–6% longer than for an equivalent horizontal dipole at the same frequency. The quarter-wavelength element lengths are found from

$$L_{meters} = \frac{76}{F_{MHz}} \text{ meters}$$

As with all antennas, the actual lengths required will be found experimentally using the calculated figure as a starting point.

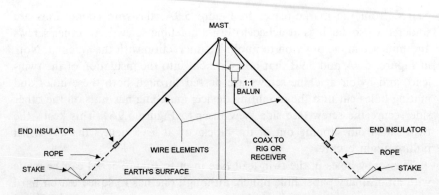

FIGURE 5.10

SLOPING DIPOLE

The **sloping dipole** (Figure 5.11) is a half wavelength dipole. Like the inverted-'vee', this dipole slants from a support (mast, tree, rooftop) to the ground. Unlike the 'vee', however, the sloping dipole has the entire half wavelength in one sloping element. The length of each quarter-wavelength element is found from the same equation as for the inverted-'vee'.

VERTICAL DIPOLE

A refinement of the sloping dipole is the vertical dipole shown in Figure 5.12. Like all verticals, the azimuthal pattern of this antenna is omnidirectional if nothing distorts it. The vertical dipole antenna works especially well in locations where there is limited space, and the desired frequencies are in the higher end of the HF spectrum. Lower-frequency vertical dipoles can be accommodated if a support of sufficient height is available. The lengths of the quarter-wavelength elements are found from $150/F_{MHz}$, although this length will almost certainly need trimming to find the actual length.

The feed shown in Figure 5.12 is a direct feed with 75 ohm coaxial cable. If you wish to substitute a 1:1 BALUN, however, feel free. Indeed, it is recommended.

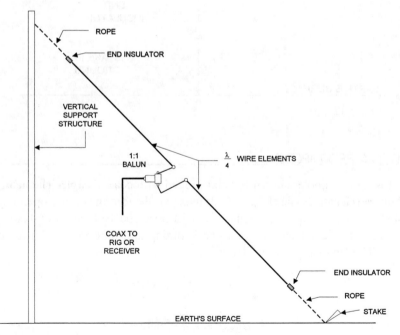

FIGURE 5.11

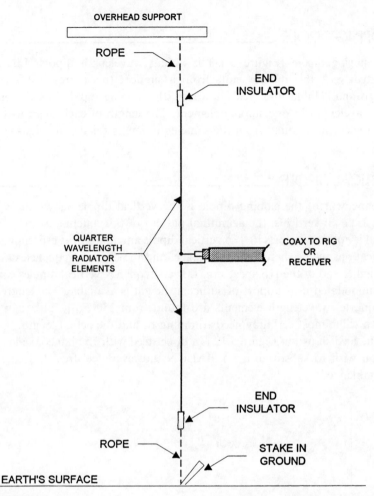

FIGURE 5.12

DELTA-FED DIPOLE

Most of the dipoles shown thus far are fed with coaxial cable. The **delta-fed half-wavelength dipole** (Figure 5.13) uses parallel feedline (either open line or 450 ohm twin-lead) and a delta match scheme. The feedline is non-resonant, and must be connected to an antenna tuning unit that has a balanced output. The lengths are found from

$$A_{\text{meters}} = \frac{142}{F_{\text{MHz}}} \text{ meters}$$

$$B_{\text{meters}} = \frac{54}{F_{\text{MHz}}} \text{ meters}$$

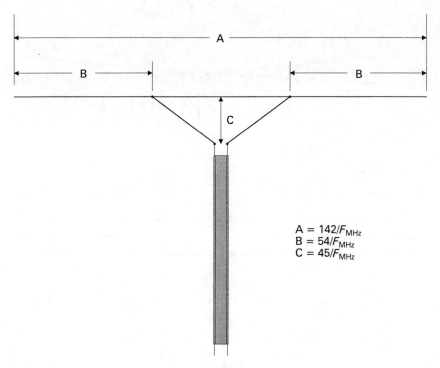

FIGURE 5.13

$$C_{\text{meters}} = \frac{45}{F_{\text{MHz}}} \text{ meters}$$

The delta-fed dipole was very popular prior to World War II, and still finds some adherents.

BOW-TIE DIPOLE

Figure 5.14 shows the 'bow-tie dipole' wide-band antenna. It is not quite like the wide-band antenna shown in Figure 5.6 because both wires in both elements are cut to the same frequency. Wide-band operation is achieved by spreading the ends of the two wires in each element approximately 11% of the total length. Length L (in meters) is found from $127/F_{\text{MHz}}$, and the spread width $W = 0.11L$. For a frequency of 4.5 MHz, therefore, the overall length of the antenna is $127/4.5 = 28.22$ m, so $W = 0.11 \times 28.22 = 3.1$ m. The bandwidth of the antenna is achieved because spreading the ends supposedly controls impedance excursions that would normally occur when the frequency departs from resonance.

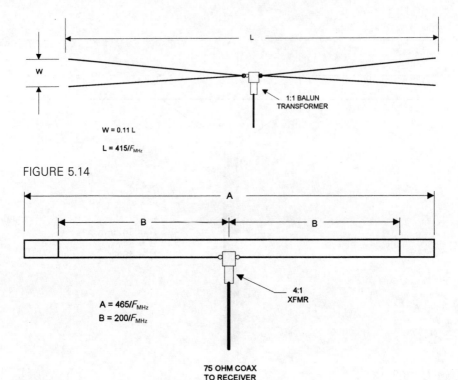

FIGURE 5.14

FIGURE 5.15

WIDER BANDWIDTH FOLDED DIPOLE

Figure 5.15 shows a wider bandwidth folded dipole antenna. This antenna is made from 300 ohm twin-lead. Like all folded dipoles, the parallel conductors are shorted at the ends, and one conductor is broken at its center to accommodate either a 300 ohm twin-lead feedline to the receiver, or a 75 ohm coaxial cable coupled to the antenna through a 4:1 BALUN transformer. In the antenna of Figure 5.15, however, there is a difference from the normal folded dipole format: shorts are placed between the two conductors at very critical distances (B) from the feedpoint. The overall length (A) of the folded dipole is given by $142/F_{MHz}$, while the distance between the inner shorts ($B-B$) is $122/F_{MHz}$, i.e. the distance from the feedpoint to each inner short is $61/F_{MHz}$.

DOUBLE EXTENDED ZEPP

The double extended Zepp antenna (Figure 5.16A) is a dipole in which the overall length is 1.28λ, and each element is 0.64λ. It provides about 3 dB

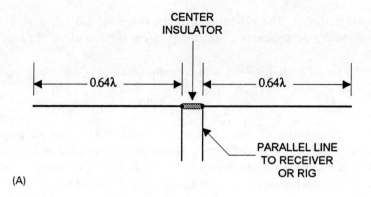

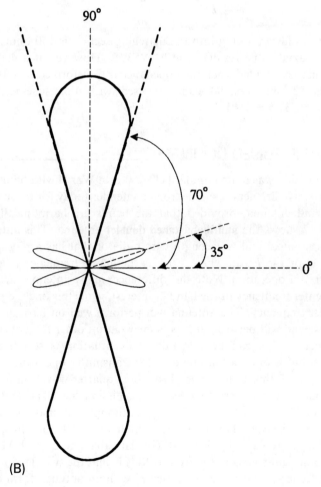

FIGURE 5.16

gain over a dipole. The azimuthal pattern (Figure 5.16B) is a narrowed figure '8' with sidelobes appearing at ±35° from the line along the antenna length.

The length of each 0.64λ element is found from

$$L_{meters} = \frac{163}{F_{MHz}} \text{ meters}$$

This antenna must be fed with parallel line connected to an antenna tuning unit with a balanced output. However, if you build a one-eighth wavelength matching section from 450 ohm twin-lead, then you can reduce the impedance to 150 ohms. The length of the 450 ohm twin-lead matching section is

$$L_{meters} = \frac{31.4}{F_{MHz}} \text{ meters}$$

The impedance looking into the matching section is 150 ohms, so using 75 ohm coaxial cable results in a 2:1 VSWR. However, if a 4:1 BALUN transformer is provided, then the impedance is transformed to $150/4 = 37.5$ ohms. If 52 ohm coaxial cable is used with this scheme, then the VSWR $= 52/37.5 = 1.4:1$.

MULTIBAND TUNED DOUBLETS

An antenna that was quite popular before World War II with ham operators and short-wave listeners is still a strong contender today for those who want a multiband antenna, provided that the bands are harmonically related. Figure 5.17 shows the **multiband tuned doublet** antenna. This antenna consists of a nominal half-wavelength wire radiator that has a physical length (in meters) of $145/F_{MHz}$, where F_{MHz} is the *lowest* frequency (expressed in megahertz) of operation. Note this difference: many antennas are designed for the center frequency of the band of interest, while this one is designed for the lowest frequency. The antenna will perform well on harmonics of this frequency, and will perform at least somewhat on other frequencies as well if a higher VSWR can be tolerated. This situation is seen for receiver operators, but is often a limiting case for transmitter operators.

The feed of this antenna is through a quarter-wavelength matching section made of 450 ohm twin-lead transmission line. The length of the quarter-wavelength section is $73/F_{MHz}$. Although one may get away with replacing the heavier 450 ohm line with lighter 300 ohm line, it is not appropriate to replace the line with coaxial cable and a 4:1 BALUN as is done in other antennas. The reason for this is that the twin-lead forms a set of tuned feeders, so is not easily replaced with an untuned form of line.

The lengths for typical antennas of this sort are given in Table 5.1.

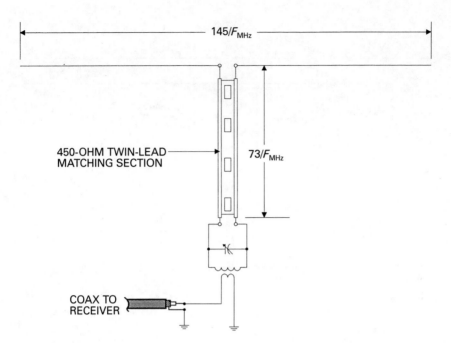

FIGURE 5.17

TABLE 5.1

Lower frequency (MHz)	Radiator element (m)	Matching section (m)
3.5	41.43	20.86
5.0	29.00	14.60
7.0	20.71	10.43

One consequence of this configuration is that a balanced antenna tuner is needed at the feed end of the matching section. The version shown here is a parallel resonant LC tank circuit that is a transformer coupled to a low-impedance link to the receiver or transmitter. The inductor and capacitor for this tuner should be designed so that they have a reactance of about 600 ohms at every frequency required. This may mean a tapped or variable inductor, or bandswitched inductor, as well as a variable capacitor. Commercial antenna tuners can also be used for this application.

THE G5RV ANTENNA

The G5RV antenna (Figure 5.18) is controversial for several reasons. One reason is the originator of the antenna. The name 'G5RV' is the call-sign of the claimed originator, Louis Varney, a British ham operator. Others claim

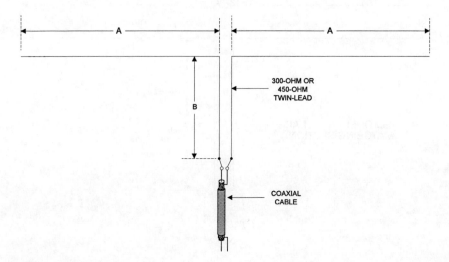

FIGURE 5.18

that the G5RV antenna is nothing but a 1930s or 1940s vintage design by Collins Radio for the US military. However, the similarities between the Collins antenna and the G5RV are, it seems to me, at best a case of 'further development' or 'co-invention,' rather than something more sinister. Because of the obvious differences between the two antennas, I prefer to continue the credit due to G5RV.

Another reason for the seeming controversy over the G5RV is perhaps the 'NIH syndrome' (not-invented-here). The G5RV is more popular in Europe than in the USA. In my own experience, the G5RV tends to be built in the US by antenna experimenters, and most of those who I have talked to are happy with the results.

Still another controversy is over whether, or how well, the G5RV works. One reviewer of one of my other antenna books (*Joe Carr's Receiving Antenna Handbook*, Universal Radio) stated that he 'wish(es) the G5RV would just go away.' The same reviewer stated that it would be better to just put up a 'dipole of the same size.' Wrong! The dipole is a single-band resonant antenna, whereas the G5RV will work on several harmonically related bands. The G5RV has two poles, but it does not exactly fit into the same category as the half-wavelength dipole.

PHYSICAL STRUCTURE OF THE G5RV

The G5RV antenna looks like a dipole, to be sure, but its length is considerably longer. Unlike many multiband antennas, the G5RV is not cut to the lowest frequency of operation, but rather to the middle frequency. For an HF ham band antenna, one designs it for 20 m (14 MHz).

Like the dipole, the G5RV is fed in the center. Unlike the dipole, a matching section made of 450 or 300 ohm twin-lead transmission line (450 ohm is preferred) is connected between the antenna feedpoint and the 75 ohm coaxial cable. The length of each radiator element (A) is

$$A = \frac{220}{F_{MHz}} \text{ meters}$$

While the length of the matching section (B) is

$$B = \frac{146 V}{F_{MHz}} \text{ meters}$$

A is the length of each radiator element, B is the length of the 450 ohm matching section, F_{MHz} is the middle frequency of operation, and V is the **velocity factor** of the twin-lead (typically 0.82 for twin-lead and 0.99 for open-wire parallel line).

In case you do not like to do arithmetic, the calculations have already been done for the HF ham bands: $A = 15.55$ meters ($2A = 31.1$ m overall), and $B = 10.37$ m for open-wire transmission line and 8.38 m for 300 or 450 ohm twin-lead. There is some argument over these figures, but they are regarded by many hams who have actually used the antenna as a good trade-off.

If you want more technical details on the G5RV, Louis Varney, the inventor, has written about the antenna in a number of publications. Varney gives the basis for operation of the G5RV antenna at 3.5, 7, 10, 14, 18, 21, 24, and 28 MHz. The VSWR on each band is a bit different, and Varney recommends the use of a transmatch or similar coaxial-to-coaxial antenna tuning unit between the transmitter and the input to the coaxial cable transmission line.

There is a possibility of an unbalanced line condition existing that causes some radiation from the feedline – and for transmitters that can cause television interference (TVI) and other forms of unpleasantness. The solution to this problem is to wind the coaxial cable into an in-line choke at the point where the coaxial cable connects to the twin-lead or parallel line matching section. This is done by winding the coaxial cable into a 15 cm diameter coil of 10 turns right at the feedpoint. The coiled coaxial cable can be secured to the center insulator by tape, string, or some other mechanism.

BALUNS ON RECEIVER ANTENNAS?

Above I mentioned a reviewer who did not like the G5RV antenna. The same reviewer stated that he did not believe that 1:1 BALUN transformers are useful on half-wavelength dipole receiving antennas. My calm, professional, well-considered response is: *rubbish*! The purpose of using a 1:1 BALUN (which after all provides no impedance transformation) is twofold. First, as in the transmit case, the BALUN prevents radiation from the transmission line. Perhaps the reviewer was thinking of TVI as the reason for avoiding feedline radiation. But, as ample test chamber evidence shows, the radiation from the feedline tends to distort the figure '8' azimuthal pattern. When a BALUN is used, the currents are balanced, and the radiation pattern is restored. And guess what? Antennas are reciprocal in nature – they work the same on receive as on transmit.

The second reason is that the receiver antenna feedline may pick up strong signals from powerful local stations. Other hams and AM broadcast band stations are particular problems. Any large signal at the input may challenge even the best receiver front-end, but if the receiver design is in any way mediocre in the dynamic range department (and many are!), then the signals picked up on the dipole transmission line shield can overload the front-end of the receiver. Using a 1:1 BALUN transformer between the antenna feedline and the radiator elements balances out the currents and seriously reduces the amount of signal seen by the receiver input.

LINEAR LOADED WIDE-BAND DOUBLET

Figure 5.19 shows an antenna that superficially resembles both the folded dipole and the G5RV, but is actually a wide-band antenna with a linear loading section between the radiators. The overall length of the main radiator elements is $146/F_{MHz}$ meters, while the linear loading element is $133/F_{MHz}$ meters long. The radiator elements are separated about 30.5 cm, with the linear loading element placed in the middle, 15 cm from each element. The linear loading element is placed horizontally about the mid-point of the antenna. The spreaders can be ceramic, or made of any insulating material. In one variant that I built, ordinary wooden dowels (1 cm diameter) were used, although I am not too sure it will weather well unless treated with some kind of coating.

This antenna is fed through 'any' length of 450 ohm twin-lead, but practical advice from several people who have actually built this antenna suggest that 'any' should be read as 'one-fifth to one-half the length of the radiator elements.' A 9:1 BALUN transformer is connected to the feedpoint, and coaxial cable routed to the receiver.

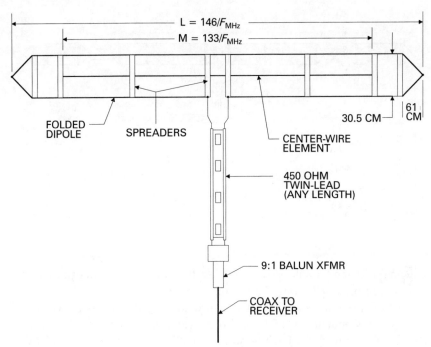

FIGURE 5.19

One practical problem with this antenna is that it tends to flop over because it is top heavy. Two solutions present themselves. First, do not tie off the ends in a single rope, but rather attach two ropes to the supporting structure at each end. Second, tie rope or fishing line from the bottom radiator at each end to a supporting structure below, or to stakes in the ground. One fellow used fishing weights for this purpose. The weights were attached to the bottom radiator element at the far ends.

THE AUSSIE BROAD-BAND HF DOUBLET FOR RECEIVERS

One of the problems with most of the antennas in this chapter is that they work on a single resonant frequency, and a narrow range of frequencies around it. Some of them will work over relatively wide bandwidths, but the coverage is nonetheless limited with respect to the entire HF spectrum. The Aussie doublet shown in Figure 5.20 is a receive antenna with a VSWR less than 2.5:1 over the entire 3–30 MHz range of the HF band.

This antenna is rather large, being 40 m long and 1.8 m wide, so requires some space to erect. Also, the two conductors of each segment are separated by insulating spacers. One approach to making these very long (1.8 m) spacers is to use PVC plumbing pipe or small size lumber. Even then, the

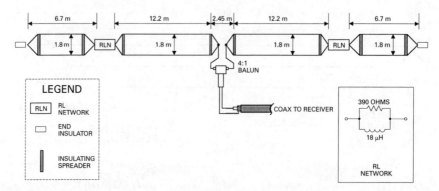

FIGURE 5.20

antenna will tend to flop over and lay flat unless rope moorings to the ground are provided along the bottom conductor.

Part of the broad-banding effect is provided by the two-wire design and their spacing, but much of it is also provided by a pair of RL networks ('RLN' in Figure 5.20). Each network is a 390 ohm resistor in parallel with an 18 µH inductor. It is these components that limit the use of this antenna for transmitting, although moderate powers can be accommodated if a suitable non-inductive, high-power resistor and a suitably designed inductor are provided.

THE CAPACITOR-TUNED WIDE-BAND DIPOLE

Dipole antennas are resonant and so will operate effectively over a relatively narrow band, as well as the third harmonic of that band. It is possible to broaden the response of the antenna by inserting a variable capacitor at the feedpoint (Figure 5.21). This capacitor should have a maximum value of 500 pF, although 730 and 1100 pF have been successfully used. The latter two values represent the values obtained when two or three sections of a 365 pF broadcast variable capacitor are used (receive only). For transmitting, a wide air gap capacitor ('transmitting variables') or vacuum capacitor should be used.

The best solution for using this antenna is to mount the capacitor at the feedpoint inside of a weatherproof housing, and then use a low-voltage, low-RPM direct-current (DC) motor to drive it. Voltage to the motor can be supplied through a separate wire, using the coaxial cable shield as the DC return line. In another scheme, it is also possible to use the coaxial cable center conductor to carry the DC (with the shield acting as the return), but it is necessary to separate the motor and signal line with an RF choke at both

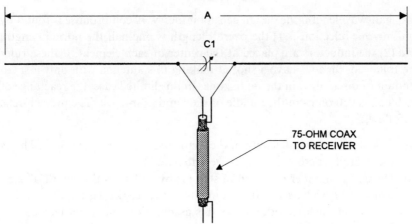

FIGURE 5.21

ends of the transmission line. IT IS VERY IMPORTANT TO USE LOW-VOLTAGE DC FOR THIS JOB. DO NOT USE AC POWER MAINS VOLTAGE!

THE INDUCTOR-LOADED SHORTENED DIPOLE

At low frequencies, dipole antennas can take up a large amount of space. At 75/80 m ham bands, for example, a half-wavelength dipole is 41 m long; at 5 MHz, the length is still 28.5 m. This fact of physics can ruin the plans of many radio enthusiasts because of space limitations. There is, however, a method for making a dipole one-half the normal size. Figure 5.22 shows a dipole that is one-half the size of the normal half-wavelength dipole, even though it simulates the electrical half wavelength.

An antenna that is too short for its operating frequency offers a capacitive reactance component to the impedance at the feedpoint. In order to tune out this impedance the opposite form of reactance, inductive reactance, is placed in series with each radiator element. Although the inductors can be placed anywhere along the line, and the overall length can be any fraction of a 'full-

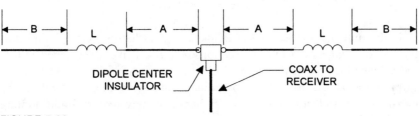

FIGURE 5.22

DOUBLETS, DIPOLES, AND OTHER HERTZIAN ANTENNAS

size' half-wavelength, the design gets a bit messy. A compromise is found by limiting our selections: (1) the overall length is one-half the normal length; and (2) the inductors are placed at the center of each element. If these rules are followed, then we have a shot at making this antenna with only a small amount of discomfort in the math zone. In this limited case, the reactance of each coil is 950 ohms in the middle of the band of interest. The procedure is as follows:

(a) Calculate the overall length of the elements: $L_{meters} = 37.5/F_{MHz}$. This is the overall length of the half-size antenna.
(b) Divide the number obtained in step (a) by 2. This is the length of each segment of wire (A or B). In our limited case, the assumption is that $A = B$, but in some other cases this assumption might not be true.
(c) Calculate the inductance required to produce a 950 ohm reactance at the design frequency:

$$L_{\mu H} = \frac{X_L \times 10^6}{2\pi F_{MHz}}$$

Antennas do not give up convenience for free, however. The price of this antenna design is narrower bandwidth.

MORE MULTIBAND DIPOLES

The shortened inductively loaded dipole of Figure 5.22 can be used on two bands by increasing the reactance of the coil to 5500 ohms at the middle frequency of the highest band. For example, you can make a 75/80 m plus 40 m dipole. The inductor in this case serves not only to foreshorten the antenna's overall length, but also acts as an RF choke to the higher-frequency band. In that case, the sections labelled A in Figure 5.22 are for the higher-frequency band, while $A + B$ is for the lower-frequency band. Approximations of the lengths are found from

$$A = \frac{74}{F_{HI(MHz)}} \text{ meters}$$

$$B = \frac{4.64}{F_{LO(MHz)}} \text{ meters}$$

In this antenna, the sections A act like a regular half-wavelength dipole at the higher frequency, while the sections $A + B$ act like a shortened inductively loaded dipole at the lower frequency. You are well advised to leave extra space on the B-sections for adjustments. This antenna tends to be narrower in bandwidth on the lower band, so may require some fiddling with lengths to achieve the desired resonance.

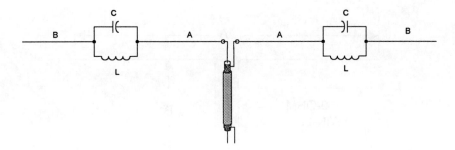

FIGURE 5.23

Another approach to multibanding is shown in Figure 5.23. This antenna is called a **trap dipole** because it uses a pair of parallel resonant LC networks to trap the higher frequency. A parallel resonant network presents a high impedance to signals at its resonant frequency, but a low impedance to frequencies removed from resonance. Like the antenna above, the sections marked A are a half-wavelength dipole at the higher frequency, while the $A + B$ sections form a near-half wavelength at the lower frequency. These lengths are approximated by the normal half-wavelength equations, although some adjustment is in order due to the effects of the trap. The traps consist of an inductor (L) and capacitor (C) in parallel. At resonance, $X_L = X_C = 370$ ohms.

MULTIBAND RESONANT ANTENNAS: MULTIBAND TUNED DOUBLET

An antenna that was quite popular before World War II is still a strong contender today for those who want a multiband antenna, provided that the bands are harmonically related. Figure 5.24 shows the multiband tuned doublet antenna. It consists of a nominal half wavelength wire radiator. The antenna will perform well on harmonics of this frequency, and will perform at least somewhat on other frequencies as well if a higher VSWR can be tolerated.

The feed of this antenna is through a matching section of 450 ohm twin-lead transmission line. Although one may get away with replacing the heavier 450 ohm line with lighter 300 ohm line, it's not appropriate to replace the line with coaxial cable and a 4:1 BALUN as is done in other antennas. The reason for this is that the twin-lead forms a set of tuned feeders, so is not easily replaced with an untuned form of line. One consequence of this configuration is that a balanced antenna tuner is needed at the feed end of the matching section. The version shown here is a parallel

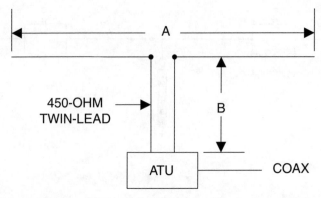

FIGURE 5.24

resonant L-C tank circuit that is transformer coupled to a low impedance link to the receiver.

The antenna is a half wavelength long at the lowest frequency of operation, and the twin-lead is a quarter wavelength long, or an odd integer multiple of quarter wavelengths. The lengths are:

$$A = \frac{144}{F_{\text{MHz}}}$$

$$B = \frac{73}{F_{\text{MHz}}}$$

This antenna depends on an antenna tuning unit (ATU) to be effective, unless the transmitter has a balanced output that can handle the high impedance. The ATU needs to be the type that has a balanced output, rather than the coaxial cable 'line flattener' variety.

THE TILTED, CENTER-FED TERMINATED, FOLDED DIPOLE

Figure 5.25 and the following figures show the *tilted, center-fed terminated, folded dipole* (TCFTFD) antenna, which is a special case of a loop antenna and a folded dipole antenna. The inventor, Navy captain G.L. Countryman (W3HH), once called it a 'squashed rhombic' antenna. The antenna is a widely spread folded dipole, and is shorter than a conventional folded dipole. It must be mounted as a sloper, with an angle from its upper vertical support of 20 to 40°.

The feeding of the antenna is conventional, with a feedpoint impedance close to 300 Ω. A 75 Ω coaxial cable is connected to the bottom half of the antenna through a BALUN transformer that has a 4:1 impedance ratio.

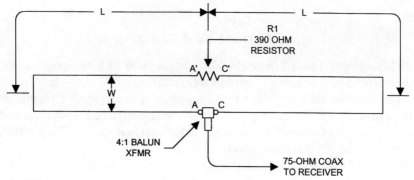

FIGURE 5.25

At the top side of the antenna, the 'feedpoint' is occupied with a termination resistor of 370 to 430 Ω (390 Ω, 1 or 2 watts, makes a good compromise for receiving antennas).

The spread (W) of the antenna wire elements is found from:

$$W = \frac{2.99}{F_{MHz}}$$

where W is the width of the antenna in meters (m) and F_{MHz} is the frequency in megahertz.

The spreaders are preferably ceramic, strong plastic, or thick-walled PVC pipe. The spreaders can be made of wood (1 × 2 stock or 1-inch dowels) for receive antennas if the wood is properly varnished against the weather.

The overall length of the antenna is calculated a little differently from most antennas. We need to calculate the lengths from the *feedpoint to the middle of the spreaders*, which is also the length from the middle of the spreaders and the terminating resistor. These lengths are found from:

$$L = \frac{54.3}{F_{MHz}}$$

where L is the length in meters (m) and F_{MHz} is the frequency in megahertz.

Four sections of wire, each with a length defined by the equation above, are needed to make this antenna.

The height of the upper antenna support is determined by trigonometry from the length of the antenna from end-to-end (not the length calculated for D, but approximately twice that length), and the angle. For example, at 7 MHz the lengths are 7.76 meters, and the spreaders are 0.47 cm. Thus, the overall physical length, counting the two element lengths and half of both spreader lengths, is [2 × 7.76 meters − (2 × 0.5)] meters, or 16.52 meters. If the angle of mounting is 30°, then the antenna forms the hypotenuse of a 60/30 right angled triangle. If we allow two meters for the lower support, then the upper support is:

$$\text{height} = 2 + L\cos Q$$
$$= 2 + (16.52\cos 60) = 10.26 \text{ meters}$$

This antenna has a low angle of radiation, and at a tilt angle of 30° (considered ideal) it is nearly omnidirectional.

The termination resistor can be mounted on a small piece of plastic, or alternatively, as shown in Figure 5.26, it can be stretched across the end insulator in the manner of the inductors in the previous section. Use a 390 Ω, 2 watt resistor for this application.

Figure 5.27 shows the installation of the antenna. The upper support and lower support are made of wood, or some other insulating material. There are ropes tying the ends of the antenna to the supports. Detail of the spreaders and end insulators are shown in Figures 5.28 and 5.29, respectively.

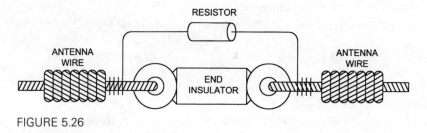

FIGURE 5.26

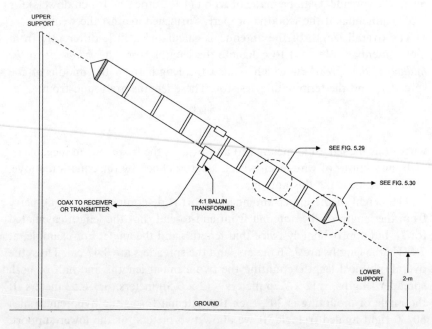

FIGURE 5.27

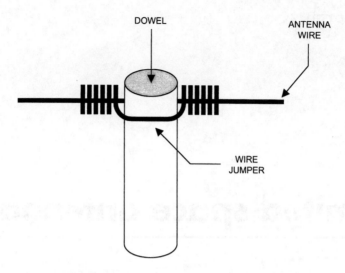

FIGURE 5.28

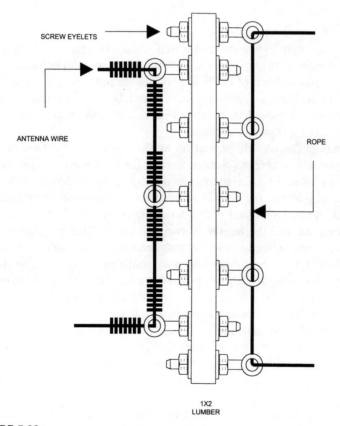

FIGURE 5.29

DOUBLETS, DIPOLES, AND OTHER HERTZIAN ANTENNAS **117**

CHAPTER 6

Limited space antennas

Sometimes at night, when things are well and all is right in the universe, I dream. I dream of owning (or at least renting from a radio enthusiast landlord) about 500 hectares of flat land good for farming. Not that I intend to do any farming (unless 'antenna farm' counts), but good farmland is moist and composed of enough conductive minerals to make a very good radio frequency (RF) ground for antennas!

Dreams are fun to indulge, but on awakening one finds the sometimes harsh nightmare of antenna construction on the limited estates that most of us can afford. And I have lived in some tight spots. In my novice days, living in my parents' home, the lot was respectable by American suburban standards, but was criss-crossed with alternating-current (AC) power lines to both our house and the houses of two neighbors. That problem sharply limited the area open to safe antenna erection. The student boarding house where I lived in college had a sympathetic landlady (who in any event was too poor of sight to notice an antenna short of a beam and tower) was on a shallow, narrow city lot with few possibilities below 20 m (and even that band strained the antenna constructor's imagination). After being married, I lived in an apartment and then a small house of my own for a number of years – all of which were limited spaces. Even with my present home, which seems a mansion compared to our first house, my ability to put up full size antennas is a testament to my wife's tolerant attitude. So what to do?

Fortunately, there are some things that can be done to accommodate the limited space situation. Nearly all of these schemes use some form of compensation antenna, and as a result do not work as well as the full-sized equivalent. The TANSTAAFL principle reigns supreme: '*there ain't no such thing as a free lunch.*' Whenever one attempts to reduce the size of an antenna for any given frequency, the result is that something else falls down a bit: gain, directivity, impedance, efficiency, bandwidth, or all of the above. While these factors may lead one to a depressed sense of gloom and doom, that is not the best attitude. The correct way to view the situation is not comparing the limited space antenna to a full size antenna, or the best in class, but rather to whether or not it allows you to operate at all. After all, a 10 m long, end-fed inductor loaded Marconi draped out of an upper-story window may not work as well as a 75 m dipole, 25 m in the air, but it works well enough to make contacts.

SOME OBVIOUS SOLUTIONS

Not all limited space antenna choices are bad. For example, one could restrict operations to the upper bands, i.e. those above 21 MHz and into the very high-frequency (VHF) region. I know one fellow who was a 300+ country DXCC award holder who operated only on the 15 m ham band because of space limitations (he lived on a small $\frac{1}{12}$ hectare lot). He was able to erect a 15 m two-element quad beam that worked quite well. Another fellow earned his DXCC as a newly licensed amateur (when operating skills are presumedly undeveloped as yet) with a 50 W kit-built transmitter and a 10–15–20 m trap vertical mounted on the roof of a four-story apartment building. It can be done.

The vertical antenna is a decent solution for many people. A full-sized 40 m (7–7.3 MHz) vertical is only 10 m high, and that is right at the legal limit or 10.7 m in my county for an installation that does not need mechanical inspection by the local government. Adding a switchable loading coil at the base of that antenna would also allow 75/80 m operation, especially if an antenna tuning unit is provided.

Modern trap verticals are also a good bet. Several commercially made verticals allow operation over the 40 m band through 10 m and are only 6.5 m tall when mounted on the ground. A good set of radials makes this antenna reasonably efficient. The only drawback is that the omnidirectional pattern makes interference rather constant. The half-wavelength bottom-fed ham-band verticals currently on the market are quite decent DX antennas.

Another obvious solution is to use any of several inductor loaded antennas such as the loops and shortened dipoles shown elsewhere in this book. Some of these antennas can be quite useful. Also, I once operated with a pair of commercial coil-loaded mobile antennas mounted back to back to form a kind of dipole. Did it work well? No! But it worked well enough for me to rack up a lot of DX using an old Heath HW-101 transceiver.

I have also used commercial mobile antennas as fixed verticals. In one case, right after we moved into our first home, I mounted a Hustler mobile antenna for 20 m on the window sill of the upstairs bedroom that served as my office (until the kids started coming along). A pair of radials out of the window, sloping to a fence, completed the 'ground' side of the antenna. It worked rather well, actually. Later on, I added 15 and 40 m coils using one of those attachments that allow three coils to coexist on the same base mast. It worked very well at 15 m and 20 m, and passably well at 40 m.

SOME OTHER SOLUTIONS

Figure 6.1 shows Marconi (Figure 6.1A) and Hertzian (Figure 6.1B) antennas for use in limited space situations. In the Marconi version, the antenna wire is mounted to lay flat on the roof, and then (if possible) to an attachment point on a support such as a mast or convenient tree. The Hertzian version is a dipole, preferably a half wavelength (if you want to use coaxial feed), positioned such that the feedpoint is in the middle of the roof line.

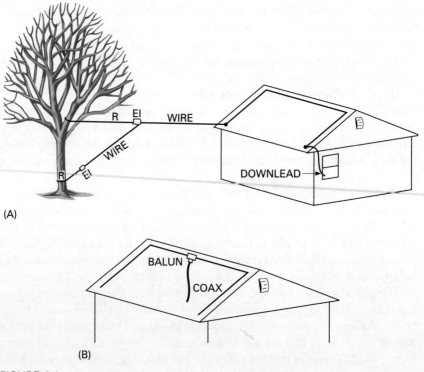

FIGURE 6.1

The idea in building these antennas is to put as much wire in the same direction as possible, but do not worry too much if the goal is not met. The real goal is to get on the air or use your receiver, depending on your interests.

Another solution is the loop antenna shown in Figure 6.2. The best loop is one that is a full wavelength on the lowest frequency of operation, but any convenient loop can be made to work at least part of the way. This is especially true if tuned parallel feeders and a decent antenna tuning unit (ATU) are used.

All of these antennas can be used either on the outside of the roof, or inside the attic. In neither case, especially the latter, would you want to try it on a roof that has a copper or aluminum base. That approach is not used on residential properties much anymore, but if you have an older dwelling the roof may well be copper covered.

Mounting the wire can be a bit of a problem, especially when the antenna is on the outside of the roof. What seems to be the best way is to use nail-in or screw-in stand-off insulators attached to the wooden roof. This works well only when you can seal it against rain water seeping in. Water will wick along the threads, and rot the wood – leading to leaks and expensive repairs. A better approach would be to design a wooden fixture attached to the soffits or overhang in a way that prevents extensive damage. Inside the attic, however, one can easily use standoffs, but they must be attached to a roof rafter (Figure 6.3), rather than the roof covering. A screw thread that penetrates to the outside surface will allow as much water damage (or more) than a thread going the other way.

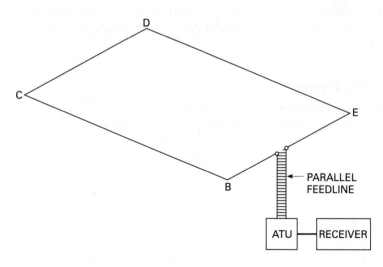

FIGURE 6.2

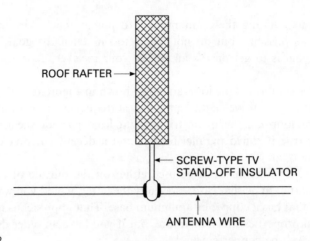

FIGURE 6.3

Marconi and Hertzian antennas may not be as good a choice as the loop because they tend to have high voltages at the ends. Use of even moderate RF power levels could cause an arcing situation that represents a fire hazard. As a result, even with the loop, if the antenna is used with a transmitter the power level should be limited.

These ideas work especially well for people with restrictive local ordinances that make decent antennas unworkable, or nasty neighbors, or restrictive covenants in their deed (and a militant homeowners association board that enforces them!). Some situations allow an antenna such as that shown in Figure 6.1A or Figure 6.1B as long as it is not visible from the street.

Zig-Zag Dipoles

The antenna purist would probably roast me over a slow fire like a marshmellow for suggesting some approaches to the limited space problem. Figure 6.4 is especially suited to drawing the ire of the pure. But it also works passably well for people with limited space. This antenna uses a zig-zag path for the dipole elements. Try to put as much wire in one direction as possible, and wherever possible make the pattern symmetrical. Although the angles shown in Figure 6.4 are acute, any angle can be used. Of course, the closer the pattern is to the regular straight dipole installation, the better it will work.

The antenna shown in Figure 6.4 is seen from the side, i.e. a ground view. The zigging and zagging can be in any direction in three-dimensional space,

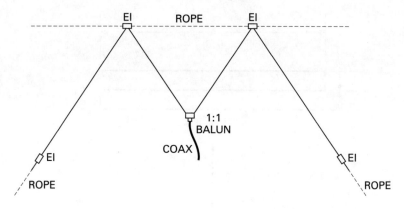

FIGURE 6.4

however. Some combinations will not work well, but others will work well enough to get you on the air.

You will find that the pattern suffers when this is done, but again the goal is to get you on the air – not to produce a perfect antenna. Also, expect to use a 'line flattening' antenna tuner to wash away the sins of the voltage standing wave ratio (VSWR).

Linearly Loaded 'Tee' Antenna

The traditional method for loading an antenna that is too short is to put a coil in series with one or all elements. In the 'tee' antenna in Figure 6.5, however, a different approach is used: **linearly loaded elements**. This antenna has been popular with low-band operators and receiver users who lack the tremendous space required for proper antennas at those frequencies. The overall length of the element (*A*) is usually around $\lambda/3$, and the spacing between the three segments (*B*) is about 25 cm (a little more or a little less can be accommodated by the design). The vertical portion of the antenna (*C*) is a quarter-wavelength single wire, and is fed at the bottom by parallel tuned transmission line.

The critical lengths are

$$A = \frac{50}{F_{\text{MHz}}} \text{ meters}$$

$$B = 25 \text{ cm}$$

$$C = \frac{75}{F_{\text{MHz}}} \text{ meters}$$

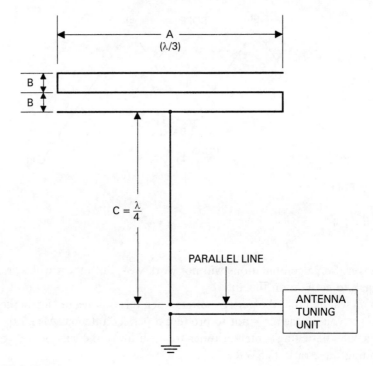

FIGURE 6.5

The Right-Angle Marconi

Figure 6.6 shows a simple quarter-wavelength Marconi antenna that can be used with 52 or 75 ohm coaxial cable. The antenna consists of two wire sections, each $\lambda/8$ long, erected at right angles to each other. The right-angle Marconi (sometimes laughingly called 'half-Hertzian') produces both a vertically and horizontally polarized signal. This method of construction reduces the linear property needed for the antenna by half, so is more accessible to a larger number of people.

Shortened Hertzian Radiators

The zig-zag dipole shown earlier is only one possibility for reducing the size of a Hertzian antenna. The antenna of Figure 6.7 is a species of zig-zag doublet, but is a little less random in its construction than the previous antenna of this sort. The overall length is $\lambda/2$, but the two $\lambda/4$ elements are bent into right-angle sections of $\lambda/12$ (vertical) and $\lambda/6$ (horizontal). The antenna is fed with open-wire parallel line operated as tuned feeders.

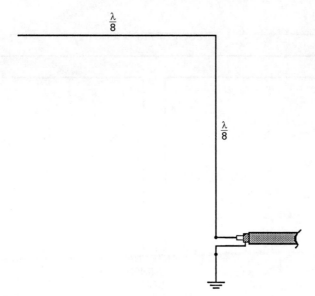

FIGURE 6.6

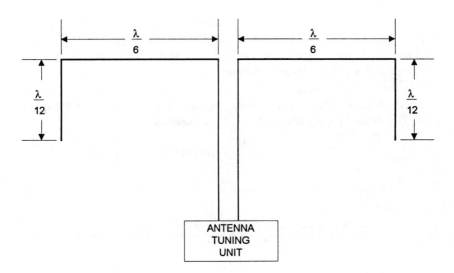

FIGURE 6.7

The antenna in Figure 6.8 looks like a folded dipole, and on one hand that is exactly what it is. The antenna length is cut for $\lambda/4$ at the lowest frequency of operation. At the second harmonic of this low band the antenna acts like a folded dipole, while at the lower frequency it is a species of 'tee' antenna. It is fed with open-wire tuned feeders. If you want to

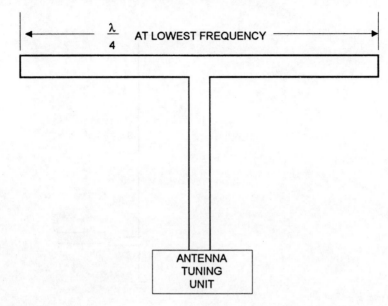

FIGURE 6.8

operate it only one band (the lower band), and wish to use coaxial cable, then a quarter-wavelength matching stub can be provided, and the antenna driven with a 4:1 BALUN transformer.

The doublet antenna in Figure 6.9 is a bit longer than $5\lambda/8$, and is fed by coaxial cable through a shorted matching stub and a 4:1 BALUN transformer. The overall length (B) of the radiator element is

$$B = \frac{102}{F_{\text{MHz}}} \text{ meters}$$

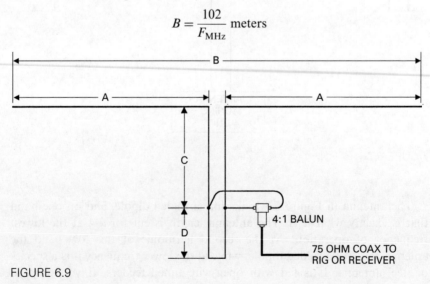

FIGURE 6.9

The matching stub consists of a length of parallel open-wire transmission line, divided into two segments either side of the feedpoint. The dimensions of the two segments are

$$C = \frac{32}{F_{MHz}} \text{ meters}$$

$$D = \frac{8.8}{F_{MHz}} \text{ meters}$$

The tap point on the stub represented by these lengths will provide a good match to 300 ohm twin-lead, or to 75 ohm coaxial cable if a 4:1 BALUN transformer is provided.

The extended folded dipole shown in Figure 6.10 uses 300 ohm twin-lead for the folded dipole section (*A*) and single conductor wire for the vertical extensions (*B*). The length overall of section *A* is found from

$$A = \frac{123}{F_{MHz}} \text{ meters}$$

While the vertical sections are

$$B = \frac{9.2}{F_{MHz}} \text{ meters}$$

The antenna of Figure 6.10 can be fed with 300 ohm twin-lead, but this requires the antenna tuning unit to have a balanced output. Alternatively, you can feed it with 75 ohm coaxial cable if a 4:1 BALUN transformer is provided at the interface between the coaxial cable and the antenna feedpoint.

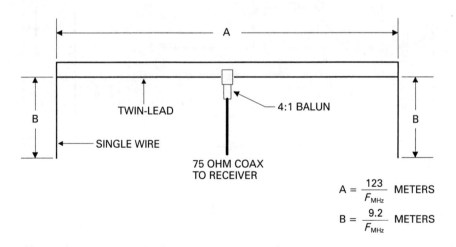

FIGURE 6.10

Limited Space Radial Layout

Radials are quarter-wavelength (usually) pieces of wire connected to the ground side of a transmission line at the antenna. These radials can be on the surface, under the surface, or elevated, depending on the particular antenna. The radials are used to form a **counterpoise ground**, i.e. an artificial ground plane. It is seen by the antenna as essentially the same as the ground.

By the way, the preferred 'ground level' configuration is buried, rather than on the surface, because it prevents injury to people passing by, either from RF burns or stumbling over the fool wire. But what do you do in the limited space situation?

A representative solution is shown in Figure 6.11. The radials shown in textbooks are straight, and that is the preferred configuration. But if you do not have the space, then you need to use some variant of the two paths shown in Figure 6.11. The radial can either go around the perimeter of the property, or zig-zag back and forth in either a triangle or rectangle pattern (the latter is shown here). I have even tacked radials to the baseboard of a student boarding house room (not recommended over a few watts of QRP power levels).

Imagination, a bit of engineering, some science, and a whole lot of luck make radio operations from limited space situations possible – or at least a lot easier.

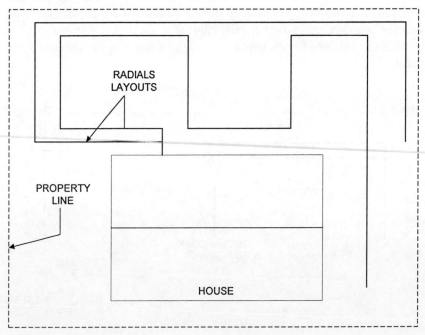

FIGURE 6.11

CHAPTER 7

Large loop antennas

There are two basic classes of loop antennas: **small loops** and **large loops**. The small loop antenna has an overall length that is less than about 0.18λ. Large loop antennas, on the other hand, have overall lengths larger than 0.5λ, and some of them are two or more wavelengths. The small loops are used for radio direction finding and for certain AM broadcast band reception problems. The topic of this chapter is the large loop antenna, of which several varieties for both transmitting and receiving are covered. We will also look at some in-between size loops that I have dubbed 'smaller large loops' to distinguish them from small radio direction finding loops.

QUAD LOOPS

The quad loop antenna (Figure 7.1) is perhaps the most effective and efficient of the large loop antennas, and it is certainly the most popular. The quad loop consists of a one wavelength loop of wire formed into a square shape. It provides about 2 dB gain over a dipole. The views in Figure 7.1 are from the horizontal perspective looking at the broad side of the loop. The azimuthal radiation pattern is a figure '8', like a dipole, with the directivity in and out of the page.

The quad loop can be fed in either of two ways. Figure 7.1A shows the feed attached to the bottom wire segment, and this produces horizontal polarization. The same polarization occurs if the feedpoint is in the top

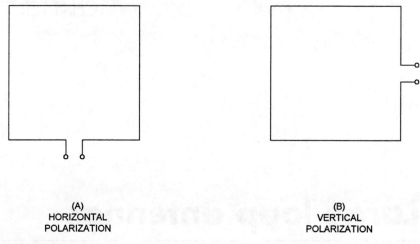

(A)
HORIZONTAL
POLARIZATION

(B)
VERTICAL
POLARIZATION

FIGURE 7.1

horizontal segment. If the feed is in either vertical segment (Figure 7.1B) then the polarization will be vertical.

The elevation pattern is shown in Figure 7.2. This pattern is found when the top horizontal segment of the loop is quarter wavelength from the ground. The directivity is in and out of the page. Note that there are two sets of lobes, one horizontally polarized (a minor lobe) and two vertically polarized. These lobes are derived by taking a slice out of the three-dimensional pattern that would be seen as a figure '8' pattern from above.

The overall length of the wire used to make the loop is found from

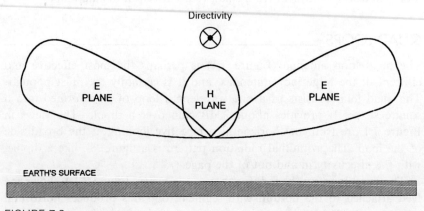

FIGURE 7.2

$$L_{\text{meters}} = \frac{306.4}{F_{\text{MHz}}} \text{ meters}$$

Examples of wire lengths for loops of various frequencies are given in Table 7.1. Each side is one-fourth of the total wire length.

There are several methods for constructing the quad loop. If you want a fixed loop, then it can be suspended from insulators and ropes from convenient support structures (tree, mast, roof of a building).

If you want to make the loop rotatable, then use a construction method similar to that of Figure 7.3A. This method uses a plywood **gusset plate**, approximately 30 cm square, to support a set of four **spreaders**. The gusset plate can be attached to a mounting pole or rotator pole with U-bolt clamps. Details of the gusset plate construction are shown in Figure 7.3B. The four spreaders are held to the gusset plate with two to four small U-bolts. The gusset plate is held to the mounting mast with two or three large U-bolts. In both cases, be sure to use a substantial size U-bolt in order to prevent breakage (use stainless steel wherever possible).

The spreaders can be made of wooden dowels at very high frequency (VHF). I have even seen larger cylindrical wooden dowels (2–3 cm diameter) used for quad loops in the upper end of the HF spectrum. For all other HF regions, however, you can buy fiberglass spreader specially manufactured for the purpose of building quad loops or quad beams. At one time, it was popular to build the quad loops from bamboo stalks. These were used as the core on which carpet was rolled, and carpet dealers would sell them for a modest price or even give them away. Today, however, carpet manufacturers use hard cardboard tubes for the roller, and these are unsuitable for building quad loops. Bamboo stalks of the right size (2.5–4 m) have all but disappeared from the marketplace.

The quad loop can be fed with coaxial cable, although it is a good idea to use a 1:1 BALUN transformer at the feedpoint if only coaxial cable is used. The impedance match is not exact, and a voltage standing wave ratio

TABLE 7.1

Center frequency (MHz)	L (overall length) (m)	L/4 (each side) (m)
3.75	81.70	20.43
5.00	61.28	15.32
7.20	42.56	10.63
9.75	31.43	7.86
14.20	21.58	5.39
21.30	14.39	3.59
24.50	12.50	3.13
29.00	10.57	2.64

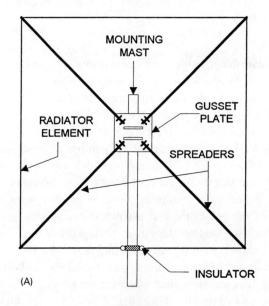
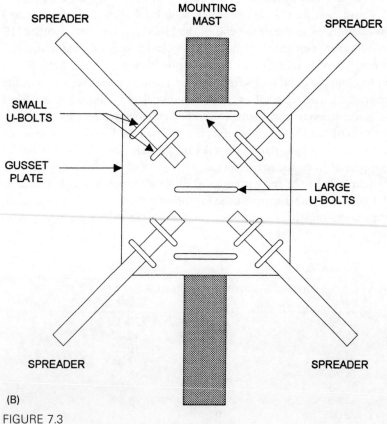

FIGURE 7.3

(VSWR) will be found. The feedpoint impedance is of the order of 100 ohms, so the VSWR when 75 ohm coaxial cable is used is only 100/75, or 1.33:1. Some people use a coaxial cable impedance matching stub between the quad loop feedpoint and the coaxial cable to the rig or receiver. Such a stub, called a **Q-section**, is made of 75 ohm coaxial cable, and is a quarter wavelength long. The coaxial cable to the rig or the receiver in that case is of 52 ohms.

To make an electrical quarter-wavelength matching stub you must shorten the physical length by the value of the velocity factor of the coaxial cable used for the matching section. For example, suppose we are building a quad loop for 14.25 MHz. A quarter wavelength in free space is $75/14.25 = 5.26$ m. But if polythene dielectric coaxial cable (velocity factor (VF) $= 0.66$) is used, the physical length required to make an electrical quarter wavelength is $5.26 \text{ m} \times 0.66 = 3.472$ m. If polyfoam dielectric coaxial cable (VF $= 0.82$) is used for the matching section, then the required physical length is 4.31 m.

QUAD LOOP BEAMS

The basic quad loop is bidirectional (figure '8' pattern). If you want to make the loop unidirectional, it can be built into a beam antenna by adding a second loop as either a **reflector** (Figure 7.4A) or **director** (Figure 7.4B). The reflector and director elements are not directly excited by the transmitter, and are therefore called **parasitic elements**. The reflector is slightly larger than the main loop (called the **driven element**), while the director is slightly smaller. The overall wire sizes are found from

Reflector:

$$L_{Ref} = \frac{315}{F_{MHz}} \text{ meters}$$

Driven element:

$$L_{DE} = \frac{306}{F_{MHz}} \text{ meters}$$

Director:

$$L_{Dir} = \frac{297}{F_{MHz}} \text{ meters}$$

Spacing between elements:

$$S = \frac{60}{F_{MHz}} \text{ meters}$$

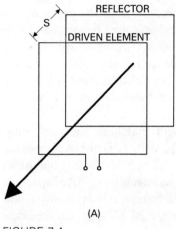
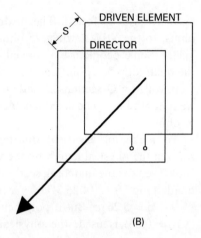

FIGURE 7.4

The directivity of the two-loop beam is always in the direction of the smallest element, as seen by the arrows in Figure 7.4, and the forward gain will be about 7 dBd (9.1 dBi), while the front-to-back ratio will be up to 25 dB (the actual value depends on the spacing of the elements).

The quad loop beam has a tradition that goes back to just before World War II. The missionary short-wave radio station HCJB in Quito, Ecuador was experiencing problems with its Yagi beam antennas. There was a constant corona discharge off the ends of the elements because the tips are high-voltage points. At lower altitudes than Quito's Andes location, these antennas do not exhibit the problem, but at altitude the arcing was severe enough to create a constant (and expensive) maintenance problem for the engineers. They invented the quad beam to solve the problem. The reason is that the feed method puts the high-voltage nodes in the middle of the vertical segments. It takes a much higher voltage to cause corona arcing from the middle of a cylindrical conductor than off the ends, so the problem was eliminated.

Ham operators and some commercial stations quickly picked up on the quad beam because it offered a relatively low-cost method for obtaining directivity and gain. In addition, the quad beam is believed to work better than the Yagi beam in installations that are close to the Earth's surface (e.g. less than a half wavelength).

The quad antennas shown here are two-element designs. Three and more elements can be accommodated by using both a reflector and a director, or any combination of multiple directors or reflectors. It is common practice (but by no means totally necessary) to use one reflector and as many directors as are needed to accomplish the desired number of elements. Each element added to the array will increase the gain and narrow directivity. The feedpoint impedance is around 60 ohms, so it makes a reasonably good impedance match to both 52 and 75 ohm coaxial cable.

DELTA LOOPS

The delta loop (Figure 7.5) gets its name from the fact that its triangular shape resembles the upper-case Greek letter *delta* (Δ). These three-sided loops are made with a full-wavelength piece of wire, each side of the equilateral triangle being one-third wavelength long.

Three different feed schemes are shown in Figure 7.5, but all three of them attach the feedline at an apex of the triangle. The antenna in Figure 7.5A is fed at the top apex, that in Figure 7.5B is fed at bottom apex, while that in Figure 7.5C is fed at one of the side apexes (either right or left could be used).

The overall length (in meters) of the wire used to make the delta loop is found from $306.4/F_{MHz}$ only if the antenna is mounted far enough from the ground and surrounding objects to simulate the elusive 'free space' ideal. But for practical delta loops a nearer approximation is found from $285/F_{MHz}$. The actual size will be between the two values.

A delta loop built for 9.75 MHz is shown in Figure 7.6. The overall formula length of the wire is 29.23 m, with each side being 9.74 m. The transmission line to the rig or receiver is 52 ohm coaxial cable. There is a coaxial quarter-wavelength impedance matching section ('Q-section') between the antenna feedpoint and the line to the receiver or rig. The Q-section is made from 75 ohm coaxial cable.

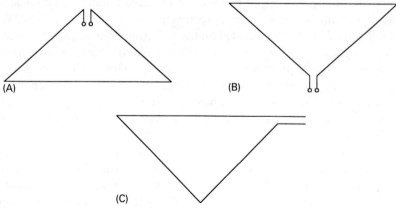

FIGURE 7.5

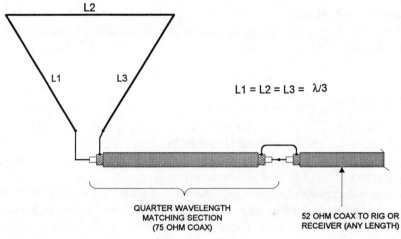

FIGURE 7.6

The physical length of the Q-section is reduced from the free space quarter wavelength ($75/F_{MHz}$ m) by the VF, to $(75 \times VF)/F_{MHz}$ m. At 9.75 MHz, if polythene coaxial cable is used (VF = 0.66) the length of the section is 5.1 m, but if polyfoam coaxial cable (VF = 0.82), then it is 6.3 m.

'ON-THE-CHEAP' ROOM LOOP

There are many situations where short-wave antennas cannot be installed without great pain and suffering, or where landlords, building managers, and other sundry unsympathetic types refuse to permit outdoor antennas. If you have a wall inside the building that will allow erection of a normal quad loop (as shown above), then there is no reason why you cannot attempt it (unless you are inside a metal frame or metal sided building!). I have seen cases where ham operators running relatively low-power levels (50–100 W) used full-wave quad loops mounted on the wall from floor to ceiling. In fact, one fellow sent me a photo of his installation where there were two adjacent quad loops on two walls that were at right angles to each other. By feeding them both along the vertical edge in the corner he was able to switch directions with a simple double-pole double-throw (DPDT) heavy duty switch. It seemed like a good solution for anyone who finds messy wire an acceptable wall decoration.

Another solution is to use a room loop antenna such as that in Figure 7.7. Although shown here as a horizontal delta loop, it can also be square or rectangular. The 'on-the-cheap' room loop is mounted on the ceiling of the room. The wire can be suspended from the type of hooks used to hang planters and flower pots. I have even seen cases where the sort of hooks

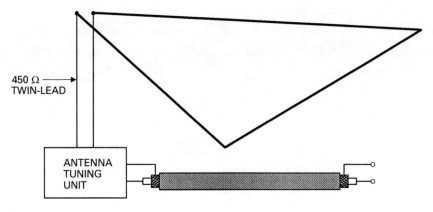

FIGURE 7.7

used to mount pictures on the walls were used on the ceiling to support the antenna wire.

The loop size is adjusted to fit the room, and is essentially a random length as far as any particular frequency is concerned. The feedline is 300 ohm, 450 ohm or parallel open feedline run vertically down the wall at the operating position to a balanced antenna tuning unit.

This antenna works well for receiving sites, and can be used for low-powered transmitters, but it should not be used for higher power transmitting. Also, it is probably prudent to use insulated wire for the antenna element.

INDOOR 'MIDDLE-EAST ROOM LOOP' ANTENNAS

There are many reasons why someone might want to build a room loop antenna. For some people, the problem is one of unsympathetic landlords, while in other situations there are problems in the layout of the property that limit antennas. In the USA, most new housing developments impose legally binding restrictive convenants that forbid the use of outdoor antennas on the deeds of houses sold. In some countries, short-wave radio antennas are limited because of political considerations. The antenna design in Figure 7.8 was sent to me via one of my magazine columns by a reader in a Middle-Eastern country that severely restricts both external antennas and the bands which short-wave receivers can pick up. He is limited to models that receive 4–7 MHz in addition to only a 200 kHz portion of the AM broadcast band and a 2 MHz segment of the FM broadcast band (what paranoia those politicians must have!). He also felt that it was necessary to not call attention to himself by erecting the short permitted outdoor antennas because, as a foreign worker, he is already somewhat suspect,

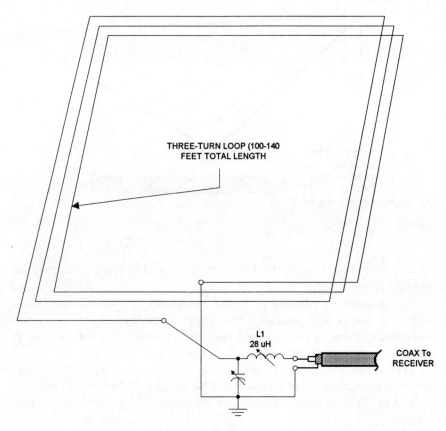

FIGURE 7.8

and believed that a substantial external antenna might label him as some sort of intelligence agent.

He claimed that this antenna worked quite well. The loop consists of 30–40 m of insulated wire (e.g. No. 22 hook-up wire) wound around the room to form three or four loops (the figure assumes a square room). My correspondent mounted the loop at the intersection of the ceiling and the walls. The windings were closely wound with little or no spacing between the windings.

The chap who sent me this design used an antenna tuner consisting of a shunt 365 pF broadcast variable and a 28 µH series inductor (i.e. a classic L-section tuner of the sort used by many radio amateurs and short-wave listeners). From some of my experiences during my student days (when I lived in student boarding house rooms), the antenna tuner might be purely optional for receiver operators. Try connecting your receiver directly to the loop using ordinary coaxial cable.

If you have some means for measuring the antenna impedance or VSWR, then make the measurement for the frequencies of interest before committing yourself to building or buying a tuner. Another option that will eliminate the antenna tuner, especially if you confine the use of the antenna to a small segment of the HF spectrum, is to use a BALUN or other form of impedance transformer at the feedpoint. This option becomes less viable as the frequency coverage required increases because the feedpoint impedance will vary all over the place with changes in frequency.

I cannot as yet recommend this antenna for transmitting above the purely QRP level (a few watts), so do not try it except at your own risk.

The directivity of this antenna is a bit uncertain because it will change with frequency. I found no problems using similar loops in my boarding house room, but common sense suggests that the azimuthal and elevation patterns will be quite different at the various bands throughout the HF short-wave bands.

A MULTIBAND SWITCHABLE DELTA LOOP

Wire antennas are among my favorites because they are low cost, reasonably well behaved, and can easily be erected and then torn down as the experimenting urge hits. Alternatively, if the urge to experiment is not something that hits you very often, or if you prudently lay down and let it pass when it does, then wire antennas are still (normally) easy to erect and are reasonably robust against the elements. They last a long time with reasonable maintenance. As might be expected of a book that deals with wire antennas, we take a look at quite a few different wire antennas. This time, we are going to look at two multiband delta loop antennas based on both the G4ABS and the N4PC designs. These antennas produce gains of the order of 4–8 dB, depending on the band of operation.

Multiband Loop 1

The first loop antenna works on 3.5, 7, 14 and 28 MHz, and is constructed of No. 14 antenna wire and a 10 m piece of either 300 or 450 ohm twin-lead transmission line. With slight modification, it will also work on a series of bands starting at 5 MHz and proceeding up through the HF short-wave band.

Figure 7.9A shows the multiband delta loop antenna design. There are four sections to this antenna. Two sloping side sections are made of No. 14 antenna wire, and are 10 m long. The top section is 9.75 m of No. 14 antenna wire, but is cut in the center to form two equal-length sections. Hanging from the center of the top section is a vertical section made of twin-lead transmission line (10 m).

To make this antenna work on the international short-wave bands rather than the ham bands, multiply these dimensions by 0.7. This is an example of dimensions scaling. It can be done by taking the ratio of the frequencies, and multiplying the resultant factor by the lengths. Scaling only works if all of the lengths are treated in this manner.

The two sloping side sections and the vertical section are brought to a connector board made of an insulating material such as ceramic, Plexiglas or even dry wood (if it can be kept dry and low power levels are used). We will take a look at the connection schemes for the different bands in a moment.

The antenna is supported in a manner similar to a dipole or any other wire antenna. The cut at the center of the top section, with its connections to the vertical section, are supported by an end insulator. The ends of the top half sections are supported by end insulators and ropes to nearby structures, just like a dipole. In some cases, a 10 m or higher wooden mast might be used to support the center of the top section. In that case, the vertical section

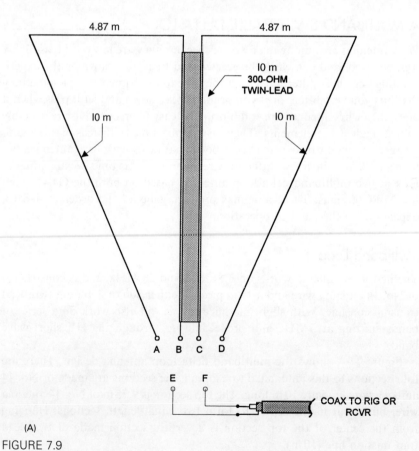

(A)
FIGURE 7.9

5		6		7		8		9		10	
414.36	375.00	497.24	450.00	580.11	525.00	662.98	600.00	745.86	675.00	828.73	750.00
214.29	197.37	257.14	236.84	300.00	276.32	342.86	315.79	385.71	355.26	428.57	394.74
107.14	104.17	128.57	125.00	150.00	145.83	171.43	166.67	192.86	187.50	214.29	208.33
74.26	73.89	89.11	88.67	103.96	103.45	118.81	118.23	133.66	133.00	148.51	147.78
53.57	52.26	64.29	62.72	75.00	73.17	85.71	83.62	96.43	94.08	107.14	104.53
41.51	41.28	49.81	49.54	58.11	57.79	66.42	66.05	74.72	74.31	83.02	82.56
35.71	34.97	42.86	41.96	50.00	48.95	57.14	55.94	64.29	62.94	71.43	69.93
30.13	30.01	36.16	36.01	42.19	42.02	48.21	48.02	54.24	54.02	60.27	60.02
26.79	25.25	32.14	30.30	37.50	35.35	42.86	40.40	48.21	45.45	53.57	50.51

435.08	357.14	522.10	428.57	609.12	500.00	696.13	571.43	783.15	642.86	870.17	714.29
225.00	187.97	270.00	225.56	315.00	263.16	360.00	300.75	405.00	338.35	450.00	375.94
112.50	99.21	135.00	119.05	157.50	138.89	180.00	158.73	202.50	178.57	225.00	198.41
77.97	70.37	93.56	84.45	109.16	98.52	124.75	112.60	140.35	126.67	155.94	140.75
56.25	49.78	67.50	59.73	78.75	69.69	90.00	79.64	101.25	89.60	112.50	99.55
43.59	39.32	52.30	47.18	61.02	55.04	69.74	62.90	78.45	70.77	87.17	78.63
37.50	33.30	45.00	39.96	52.50	46.62	60.00	53.28	67.50	59.94	75.00	66.60
31.64	28.58	37.97	34.30	44.29	40.02	50.62	45.73	56.95	51.45	63.28	57.17
28.13	24.05	33.75	28.86	39.38	33.67	45.00	38.48	50.63	43.29	56.25	48.10

OK	OK	OK	OK	OK
OK	OK	OK	OK	OK
OK	OK	OK	OK	OK
OK	OK	OK	OK	OK
OK	OK	OK	OK	OK
OK	OK	OK	OK	OK
OK	OK	OK	OK	OK
OK	OK	OK	OK	OK
OK	OK	OK	OK	OK

4.81	5.63
5.72	6.33
6.66	7.5
7.86	8.72
9.62	11.25
11.43	12.66
13.32	22.5
22.87	45
45.73	56.25

0 — 4.81
5.63 — 5.72
6.33 — 6.66
7.5 — 7.86
8.72 — 9.62
11.25 — 11.43
12.66 — 13.32
22.5 — 22.87
45 — 45.73

Input length in metres: 13

Bottom	Top	1		2		3		4		
2.000	1.810	5.05	5.36	10.10	10.71	15.15	16.07	20.20	21.43	OK
3.500	3.800	6.00	6.03	12.00	12.05	18.01	18.08	24.01	24.11	OK
7.000	7.200	6.99	7.14	13.99	14.29	20.98	21.43	27.97	28.57	OK
10.150	10.100	8.26	8.30	16.51	16.60	24.77	24.91	33.03	33.21	OK
14.350	14.000	10.45	10.71	20.91	21.43	31.36	32.14	41.87	42.86	OK
18.068	18.168	14.78	14.85	29.56	29.70	44.33	44.55	59.11	59.41	OK
21.450	21.000	20.83	21.43	41.67	42.86	62.50	64.29	83.33	85.71	OK
24.890	24.990	39.47	42.86	78.95	85.71	118.42	128.57	157.89	171.43	OK
28.000	29.700	75.00	82.87	150.00	165.75	225.00	248.62	300.00	331.49	OK

Add 5% allowance

1		2		3		4	
4.81	5.63	9.62	11.25	14.43	16.88	19.24	22.50
5.72	6.33	11.43	12.66	17.15	18.98	22.87	25.31
6.66	7.50	13.32	15.00	19.98	22.50	26.64	30.00
7.86	8.72	15.73	17.43	23.59	26.15	31.45	34.87
9.96	11.25	19.91	22.50	29.87	33.75	39.82	45.00
14.07	15.59	28.15	31.19	42.22	46.78	56.30	62.38
19.84	22.50	39.68	45.00	59.52	67.50	79.37	90.06
37.59	45.00	75.19	90.00	112.78	135.00	150.00	180.00
71.43	87.02	142.86	174.03	214.29	261.05	285.71	348.07

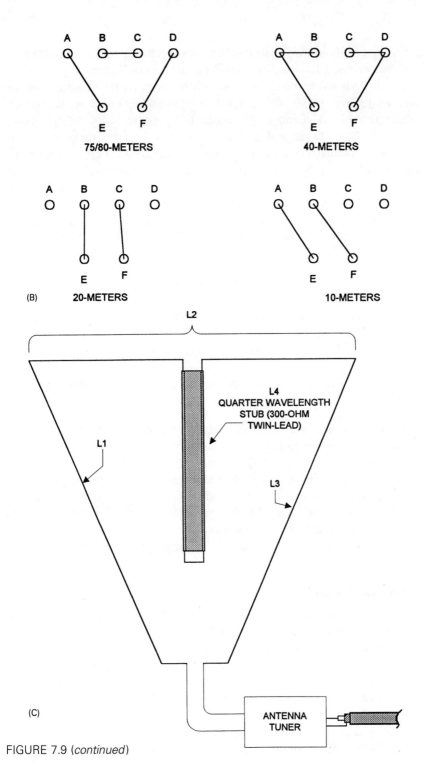

FIGURE 7.9 (*continued*)

LARGE LOOP ANTENNAS

can be tied to it, improving its mechanical support. In addition, the connections board (used for band changing) can be mounted close to its base.

Figure 7.9B shows the connections scheme to make the antenna work on the four different bands (which, you will undoubtedly notice are the harmonically related HF bands). All connections go to stand-off insulators, labeled A, B, C, D, E, and F. The cheapest way to use these connection points is to manually connect a shorting wire between the appropriate terminals. People with a bit more money to spend might want to get some coaxial relays, radio frequency relays, or vacuum relays in order to perform the switching. For receiver and very low-power transmitter applications, the connectors can be five-way binding posts that accept a banana plug. The shorting could be done with short (very short!) pieces of wire tipped at each end with the banana plugs. The connections scheme is as follows:

(1) To make the antenna work on the 75/80 m bands (3.5–4.0 MHz), short B–C, A–E, and D–F.
(2) To make the antenna work on 40 m (7.0–7.3 MHz), short B–A–E and C–D–F.
(3) To make the antenna work on 20 m (14.0–14.35 MHz), short B–E and C–F.
(4) To make the antenna work on 10 m (28.0–29.7 MHz) short A–E and B–F.

My impression of the multiband delta loop is that it works about as well as a dipole, and is generally bidirectional. However, the pattern does seem to change considerably from band to band (which is to be expected). I do not want to say much about how it changes because propagation effects are so different on the four bands that it would take a very long time of observing signals to get a good idea of what the pattern looks like. Any statement about patterns, other than it appears to change from band to band, would be unwise without extensive modeling or some actual measurements.

Multiband Loop 2

The other form of multiband loop is shown in Figure 7.9C. This design uses a full-wave loop with the balanced feedline connected at one of the apexes of the triangle, and a quarter-wavelength stub. Both the stub length and the overall length (L + L2 + L3) are calculated at the *highest* frequency of operation. It will operate over three bands that are harmonically related (e.g. 3.5/7.0/14.0 MHz, or 5/10/15 MHz, etc.). The overall length in meters is found from $(L1 + L2 + L3) = 615/F_{MHz}$, while the stub is $60/F_{MHz}$ (assuming that the usual form of twin-lead is used, which has a VF of 0.82). Both of these antennas produce patterns broadside from the loop plane, although both the azimuthal and elevation radiation patterns vary from one band to another.

DOUBLE-DELTA LOOP

The double-delta loop antenna (Figure 7.10) consists of a back-to-back pair of large horizontal delta loops fed 180° out of phase with each other. The view shown is a plan view, i.e. as seen from above. The far corners of the two delta loops are connected together by a 180° phase-reversing harness made from a length of 450 ohm twin-lead or parallel ladder line. The 180° phase reversal is achieved by twisting the twin-lead over on itself once (and only once).

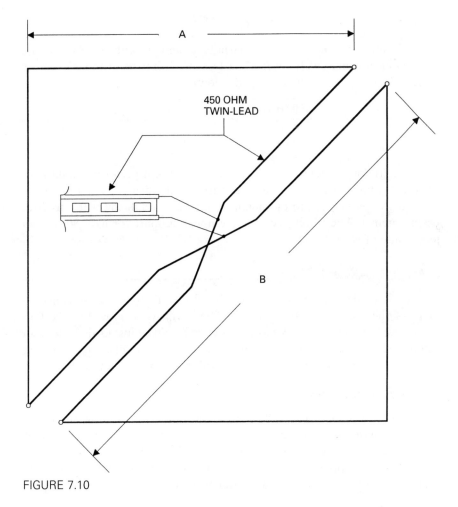

FIGURE 7.10

The feedline to the receiver or its antenna coupler is another piece of 450 ohm twin-lead connected to the mid-points of the two conductors of the 180° phasing harness. This transmission line will make a decent if not perfect match to receivers that have a balanced input. Use of a 4:1 BALUN transformer at the feedpoint will drop the impedance to about 112 ohms, which is not too bad a match for 75 ohm coaxial cable. On the other hand, if you want a near-perfect match, then opt instead for a 9:1 BALUN transformer and connect the receiver using 52 ohm coaxial cable. There are several forms of 9:1 BALUN transformer on the market, but are sometimes hard to find except by mail order. If such a transformer is not available, then you will have to build it yourself (which is not too great a chore).

The length of each of the four sides of the double-delta loop is about a quarter wavelength, and is found from

$$A = \frac{66}{F_{MHz}} \text{ meters}$$

Let us look at some examples: a 60 m band antenna cut for 5 MHz and a 31 m band antenna cut for 9.75 MHz (assume I did the arithmetic on my calculator, with the results tabulated below):

F (MHz)	Meters
5.00	13.20
9.75	6.77

The delta loop antenna is well regarded because it performs similarly to the quad loop, but is somewhat easier to install when the site only has a single high point. A friend of mine in Ireland uses a tall tower for his 20 m beam antenna. The tower becomes the upper support for a 75/80 m delta loop made of wire.

SMALLER LARGE LOOPS

The loops in this section are physically smaller than the one-wavelength loops of the previous sections. Few of them will work as well as the full-size loops, but they will work passably well, and can be installed at sites that will not accommodate the larger variety. I call them 'smaller large loops' to distinguish them from the small loops used in radio direction finding.

Half-wavelength loop

The half-wavelength loop is $\lambda/8$ on each side, and is one-half the size of the full-wavelength quad loop discussed earlier. There are two configurations for this loop. The closed-loop version (Figure 7.11A) transmits and receives in the direction opposite the feedpoint, while the open-loop variety (Figure 7.11B) transmits and receives in the opppposite direction.

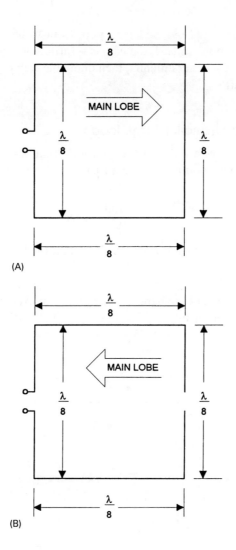

FIGURE 7.11

The two half-wavelength loops differ not just in the direction of radiation but also in the feedpoint impedance. The antenna in Figure 7.11A (closed-loop version) is fed at a voltage node, while the current node is in the middle of the opposite segment. As a result, the feedpoint impedance is several kilo-ohms. This means that a matching stub, impedance transformer or some other means of reducing the impedance is necessary. The open-loop variety (Figure 7.11B) forces the current to be a minimum at the open-circuit point, which means that the current maximum (and voltage minimum) occurs at the feedpoint. The open-loop version can be directly fed with coaxial cable.

These antennas are lossy, so have a poor front-to-back ratio (≈6 dB) and no gain over a dipole. Indeed, the gain is −1 dBd. Of course, if you want to make it seem like a better antenna, then quote the gain over isotropic, which is about +1.1 dBi.

Inductor-loaded smaller large loop

Figure 7.12A shows a less-than-full-size loop that is inductively loaded to lower the resonant point. The overall length of the wire used to form the loop is found from

$$A = \frac{180}{F_{\text{MHz}}} \text{ meters}$$

Each side is one-fourth this length. Note that the loop is a bit larger than a half wavelength.

The inductor should have a reactance of about 2500 ohms at the frequency in the middle of the band of operation. For example, if you build the antenna for reception in the 31 m band (e.g. 9.75 MHz), then the overall length of the antenna is 180/9.75 = 18.46 m, and each side is 4.62 m long.

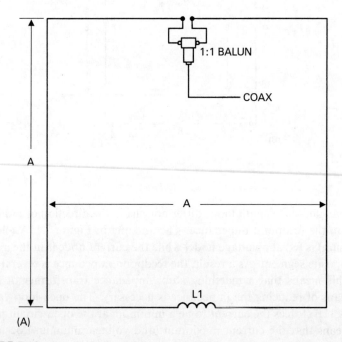

(A)

FIGURE 7.12

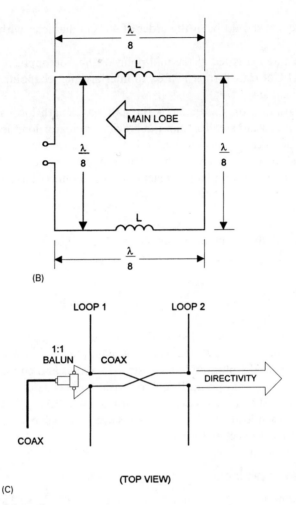

FIGURE 7.12 (continued)

The coil inductance required can be found from:

$$L_{\mu H} = \frac{400}{F_{MHz}} \text{ microhenrys}$$

For the 9.75 MHz example antenna, the value of the inductance required is (400/9.75 MHz) = 41 µH.

For receiver sites, the inductor can be a small toroid core coil. For example, if the 2.5 cm red core T-106-2 is used, then 55 turns of small diameter wire are required to make the inductor.

The antenna is not recommended for transmitting above the lowest power levels. Even at modest power levels (e.g. 100 W), a solenoid wound air core coil should be of the order of 4–5 cm diameter with wire in the No. 14 size.

The antenna is fed at the mid-point of the side opposite the inductor. A 1:1 BALUN transformer is used to interface the feedpoint to 52 or 75 ohm coaxial cable.

A variation on the theme is the half-wavelength loop shown in Figure 7.12B. It is a tad smaller than the previous inductor load loop. In this case, the inductance is split into two inductors, and are placed at the mid-points of the segments adjacent to the segment containing the feedpoint. The coils in this variety should have a reactance of 360 ohms, making the inductance

$$L_{\mu H} = \frac{57}{F_{MHz}} \text{ microhenrys}$$

Both of these antennas can be used in a beam array by placing two identical loops facing each other (Figure 12.7C), and separated by a spacing of

$$S_{meters} = \frac{33}{F_{MHz}} \text{ meters}$$

The two antennas are fed in-phase, so there are two methods of attaching the transmission line. If you want to use parallel line or 450 ohm twin-lead, then connect the phasing harness straight across and connect the transmission line to the mid-point. But if you wish to use coaxial cable, then feed the antenna as shown in Figure 12.7C, using a 1:1 BALUN transformer. The 450 ohm twin-lead phasing harness between the elements should be twisted over on itself once, as shown.

The diamond loop

The diamond loop antenna is shown in Figure 7.13. This antenna is a shortened, flattened quad loop consisting of two equilateral triangles back to back. The length of each side, plus the height (dashed line) of the two principal apexes, is found from

$$L_{meters} = \frac{70}{F_{MHz}} \text{ meters}$$

In order to feed this antenna with coaxial cable, a 4:1 BALUN transformer is used between the coaxial cable and the feedpoint of the antenna.

Half-delta loop

The antenna in Figure 7.14 achieves a smaller size by being grounded. It is essentially one-half of a full-wave delta loop, with the other half being 'imaged' in the ground. The wire forms a right triangle, with the vertical

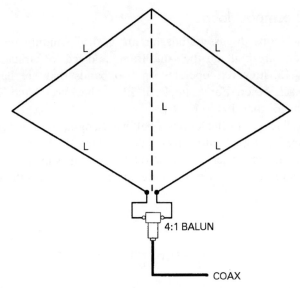

FIGURE 7.13

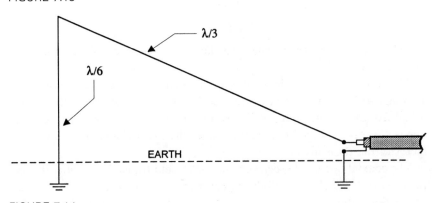

FIGURE 7.14

section being $\lambda/6$ and the sloping section being $\lambda/3$. These lengths are found from

$$\frac{\lambda}{3} = \frac{102}{F_{\text{MHz}}} \text{ meters}$$

and

$$\frac{\lambda}{6} = \frac{61}{F_{\text{MHz}}} \text{ meters}$$

The antenna is inherently unbalanced with respect to ground, so can be fed directly with 52 ohm coaxial cable.

Two-band compact loop

Most of the loops discussed thus far are basically monobanders, unless multiple loops are built on the same frame and fed in parallel. The loop in Figure 7.15, however, operates on two bands that are harmonically related to each other. For example, if FL is the lower band, and FH is the higher band, then FH = 2 × FL.

The overall length of the loop is a half wavelength, but it is arranged not into a square but rather into a rectangle in which the horizontal sides are twice as long as the vertical sides, i.e. the horizontal elements are a quarter wavelength and the vertical sections are one-eighth wavelength. The section lengths are

Horizontal:

$$L_{\text{hor}} = \frac{75}{F_{\text{MHz}}} \text{ meters}$$

Vertical:

$$L_{\text{ver}} = \frac{37.5}{F_{\text{MHz}}} \text{ meters}$$

In both equations, the frequency is the center frequency of the lower band of operation.

The vertical stub is made of 600 ohm parallel open-wire transmission line, although 450 ohm twin-lead could also be used. The length of the stub is found from the same equation (above) as the vertical segment if open-wire line is used. If twin-lead is used, then multiply that distance by the velocity factor of the transmission line.

The coaxial line is a Q-section made of 75 ohm transmission line. It is cut to a quarter wavelength of the upper band, i.e. twice F_{MHz} used in the calculations above.

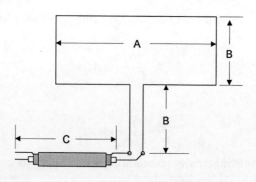

FIGURE 7.15

THE BI-SQUARE REALLY LARGE LOOP

The bi-square loop in Figure 7.16 is twice as large as the quad loop discussed earlier in this chapter. The overall length of the wire is two wavelengths, so each side is a half wavelength long. The overall length is calculated from

$$L_{meters} = \frac{585}{F_{MHz}} \text{ meters}$$

While each side is:

$$L_{meters} = \frac{146.25}{F_{MHz}} \text{ meters}$$

The bi-square antenna can be used on its design frequency, and also at one-half of its design frequency (although the patterns change). At the design frequency, the azimuthal pattern is a clover leaf perpendicular to the plane of the loop, and is horizontally polarized. At one-half the design frequency the radiation is vertically polarized, and the directivity is end fire.

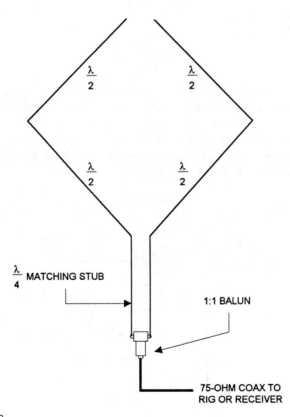

FIGURE 7.16

LARGE LOOP ANTENNAS **151**

VHF/UHF QUAD BEAM ANTENNA

Figure 7.17 shows a VHF/UHF quad beam with four elements. Although it's technically possible to build such a beam for a lower frequency than VHF, it is not too practical because of mechanical considerations. But at VHF and UHF frequencies the dimensions of the elements become less large, and therefore more manageable. These dimensions are:

Reflector:
$$L_{REF} = \frac{31\,500}{F_{MHz}} \text{ cm}$$

Driven element:
$$L_{DE} = \frac{30\,600}{F_{MHz}} \text{ cm}$$

Directors:
$$L_{DIR} = \frac{29\,700}{F_{MHz}} \text{ cm}$$

Spacing between elements:
$$S = \frac{6000}{F_{MHz}} \text{ cm}$$

Where all elements lengths are in centimeters (cm).

The antenna can be made of thin or thick wire. In the thin wire (#10 through #18) version, a support such as Figure 7.3 is needed. In the thick wire version, clothes line or other thick wire (#00 to #10) is needed. In that case, it might be possible to erect the antenna without any support other than the boom that holds all four elements.

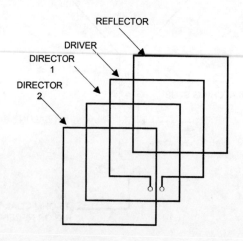

FIGURE 7.17

CHAPTER 8

Wire array antennas

The antennas found in this chapter are any of several array designs that use quarter-wavelength, half-wavelength, or other radiator elements arranged in a fixed array. There are several ways to categorize array antennas. A **collinear array** consists of two or more elements, ranging in length from 0.25λ to 0.65λ (some texts say 1λ), fed such that the currents in the elements are *in-phase* with each other.

Arrays are also classified according to the directivity: **broadside arrays** radiate and receive along a line perpendicular to the plane of the antenna, while **end-fire array**s radiat and receive off the ends of the elements.

The down side of these types of antennas is that many of them require parallel open-line transmission line (although some can use 300 or 450 ohm twin-lead), and impedance matching stubs. But we can handle that problem. Chapter 3 showed what parallel line and the two forms of twin-lead transmission line look like.

Some of the antennas can be fed directly with a 4:1 or 1:1 BALUN transformer and 75 ohm coaxial cable, but in many cases an **impedance matching stub** is needed (see Chapter 11). The BALUN transformers can either be homemade, or store-bought. Although antenna and scanner supplies distributors sell very high-frequency (VHF) BALUN transformers, you can also use television-style BALUN transformers. These may not be called BALUNs on the package, but are recognized by the fact that they have a coaxial 'F' connector at one end, and a piece of 300 ohm twin-lead at the other.

153

There are several antennas that fit either tightly or loosely into the general category of 'array antennas.' The selection of which antennas to put in this chapter, and which to put in Chapter 5 (dipoles, doublets), was not entirely clear in each case. This ambiguity was especially annoying when looking at antennas such as the double extended Zepp, which I put in Chapter 5. The problem is that antennas such as the double extended Zepp are quite properly called 'array' antennas even though they look suspiciously like doublets (which they are, of course). Other antennas were less difficult to place unambiguously in one chapter or the other.

ARRAY ANTENNAS

The idea in an array antenna is to provide directivity and gain by using two or more antenna elements in such a way that their fields combine and interact to focus the signal in one direction, or a limited number of directions (e.g. bidirectional, like a dipole). Figure 8.1A shows an array not unlike some of those in this chapter. It consists of two half-wavelength dipoles stacked one on top of another a distance (S) apart. The view here is broadside from the horizontal direction, so the dipole fed at A1–A2 is positioned above that which is fed at B1–B2.

One requirement of this form of antenna is that the two half-wavelength dipoles be driven in-phase with each other (some other antenna arrays want out-of-phase excitation). That means that the phase of the signal applied to A1–A2 is exactly the same as the phase of the signal applied to B1–B2. This is accomplished with either of the two methods shown in Figure 8.1B and

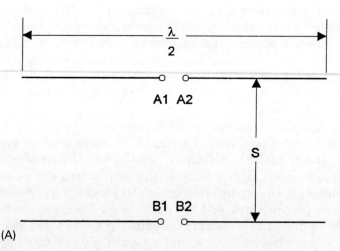

FIGURE 8.1

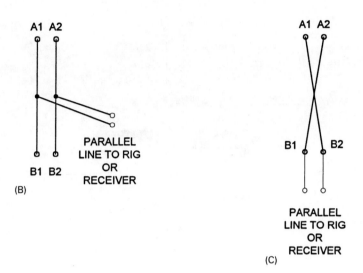

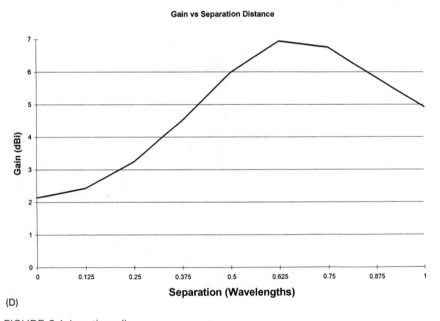

(D)

FIGURE 8.1 (continued)

8.1C. In Figure 8.1B, a piece of parallel transmission line is connected between the feedpoints such that A1 is connected to B1, and A2 is connected to B2. Another piece of parallel feedline from the receiver or transmitter is connected at the exact mid-point on the harness between the two dipoles.

(E)

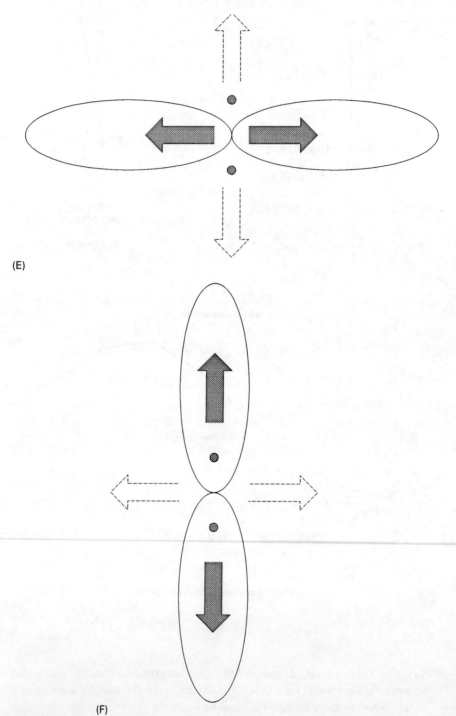

(F)
FIGURE 8.1 (*continued*)

When this is done correctly, the voltages appearing across A1–A2 are exactly the same amplitude and phase as the voltages across B1–B2.

In the method used in Figure 8.1C, the parallel line from the rig or receiver is connected to the feedpoint of one dipole (in this case B1–B2). In order to preserve the 180° phase relationship between the elements, the feedline phasing harness between A1–A2 and B1–B2 must be twisted once such that A1 is connected to B2, and A2 is connected to B1.

The gain realized by stacking the two half-wavelength dipoles in this manner is a function of the spacing (S) between the elements. The approximate gain above isotropic that can be realized is shown in Figure 8.1D (note: to convert to gain above a dipole, subtract 2.14 dB from each value). Note that the gain peaks at about 6.94 dBi (or 4.8 dBd) in the vicinity of $5\lambda/8$ (0.625λ). This is the maximum gain realizable from this configuration. Although it looks good, there are sometimes other factors that deteriorate gain performance, so some experimentation with element spacing is usually in order for those who desire peak performance.

The antenna pattern produced by phased array antennas depends on the nature of the elements, the spacing between them and the ratio of the currents flowing in the elements. In the simple case of two elements spaced a half wavelength apart, we will see the general patterns of Figures 8.1E and 8.1F. These patterns are azimuthal plots as seen from above; the antennas are omnidirectional vertical radiators. The pattern in Figure 8.1E is for the case where the two antenna elements are fed in-phase with each other. The maxima are perpendicular to a line between the two antennas, while the minima are off the ends. Feeding the antennas out-of-phase with each other (Figure 8.1F) flops the pattern over so that the maxima are along the line between the elements, and the minima are perpendicular to this line.

There are a large number of patterns for various spacings and phase angles. The originals were developed in 1929 by a chap named Brown, and published in the USA in *The Proceedings of the Institute of Radio Engineers* (now the Institute of Electrical and Electronics Engineers). Brown's patterns are reproduced in *The ARRL Antenna Handbook* and any of several antenna engineering handbooks. Now let us look at some antennas.

TWO-ELEMENT COLLINEAR ARRAY

One of the simplest forms of gain array antenna is the two-element collinear array (shown in Figure 8.2 as a side view). It consists of two half-wavelength dipoles spaced a half wavelength apart. This antenna gives a gain of 3 dB, and is a broadside array (when viewed from above the pattern is perpendicular to the line between the antenna elements).

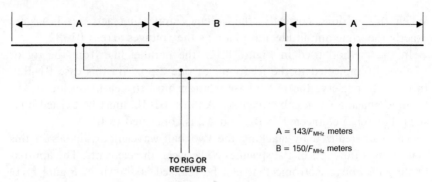

FIGURE 8.2

Each antenna element is a half wavelength, but the actual lengths and the equations used to calculate them are a little different for the HF and VHF/UHF bands.

HF band lengths

On the HF bands (3–30 MHz) the antenna is most likely mounted close to the Earth's surface, so there will be some end-effect reduction of the antenna length due to capacitive coupling. There is also a small reduction (2–3%) due to the velocity factor of the signal in the wire (this is set by the length/diameter ratio of the wire). As a result, the length of the elements is typically reduced 4–5% compared with the free space length. The gap between the dipoles, however, is calculated at the free space value. Thus,

$$A = \frac{143}{F_{MHz}} \text{ meters}$$

and

$$B = \frac{150}{F_{MHz}} \text{ meters}$$

Typical lengths for popular bands are given in Table 8.1.

TABLE 8.1

Band (MHz)	A (m)	B (m)
7.00	20.40	21.43
9.75	14.63	15.40
14.20	10.06	10.55
16.00	8.93	9.39
21.25	6.70	7.07
24.40	5.85	6.16

158 ANTENNA TOOLKIT

VHF/UHF band lengths

The physical length of the dipoles for VHF can be calculated from the free space version of the equation because these antennas are typically installed a sufficient number of wavelengths above the Earth's surface that the free space constant is reasonable. Thus, both A and B are the same physical length. These lengths are found as $A_{cm} = B_{cm} = 2324/F_{MHz}$, where F_{MHz} is the frequency in megahertz. Thus, at 150 MHz, the half wavelength is 15.49 cm; at 450 MHz it is 5.16 cm; and at 800 MHz it is 2.91 cm.

The antenna in Figure 8.2 is a little inconvenient in one respect: it requires two transmission lines connected in parallel. This is easily accomplished by using 75 ohm coaxial cable from each antenna element to a 'tee' connector, and then a piece of 52 ohm cable from the 'tee' connector to the receiver. At HF frequencies this antenna can take up a lot of 'footprint' space.

A different solution is shown in Figure 8.3. The antenna elements are separated by only a few centimeters, rather than a half wavelength. The gain drops to 1.6 dB for two elements (although for three elements it is 3 dB, and for four elements it is 4.2 dB). The antenna is fed by 450–600 ohm parallel line or twin-lead. The feedpoint impedance will be as low as 1 kilo-ohm if tubing is used, and up to 4–6 kilo-ohms if wire is used. This is one of those antennas where a matching stub is needed.

The four-element version of the antenna is shown in Figure 8.4 (it is called the **collinear array** or **Franklin array**). This antenna is similar to the two-element version of Figure 8.3, but with an additional half-wavelength element tacked onto the ends. The additional elements are separated by quarter-wavelength **phase reversal stubs**. The stubs are needed to keep the phase of the currents flowing in the outer elements the same as the currents flowing in the center elements. These stubs are made from the same type of

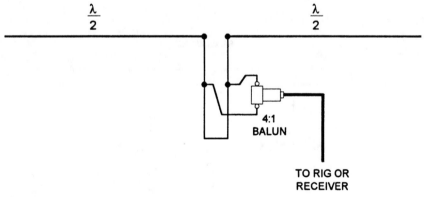

FIGURE 8.3

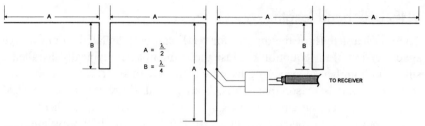

FIGURE 8.4

transmission line as the half-wavelength matching stubs, but are half as long.

Another collinear array is shown in Figure 8.5. This antenna offers 3 dB of gain, and can be fed directly from 75 ohm coaxial cable through a 4:1 BALUN transformer. The antenna consists of three half-wavelength elements, but the center element is fed at the center point (a quarter wavelength from each end). The quarter-wavelength phase reversal stubs are used between the center element and the two outer elements.

FOUR-ELEMENT BROADSIDE ARRAY

A four-element vertical broadside array is shown in Figure 8.6. Although often implemented using aluminum tubing, the array can easily be erected using wire elements as well. This antenna has four half-wavelength elements spaced a half wavelength from each other. Like the antenna above, the radiation direction is perpendicular to the line between the antennas, or in and out of the page.

The antenna elements are fed by a length of 600 ohm parallel line that connects all four together. But notice that the outside elements are fed by reversing the phase. This means that the transmission line must be twisted

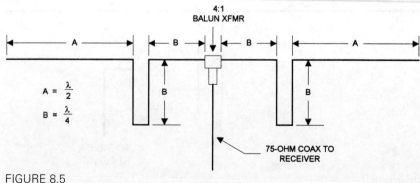

FIGURE 8.5

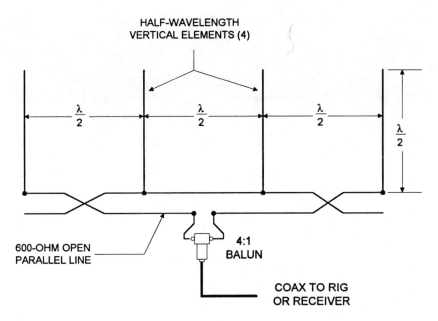

FIGURE 8.6

before being connected to the end elements; the unused conductor is left floating. The feedpoint impedance of this antenna is 200–300 ohms, so it can be fed with either 52 or 75 ohm coaxial cable if a 4:1 BALUN transformer is provided. Gains of the order of 6–8 dB are realized with this antenna.

EIGHT-ELEMENT BROADSIDE ARRAY

The eight-element broadside array shown in Figure 8.7 is built using similar methods to the previous antenna, but with two bays of elements rather than just one. The overall height of the array is one wavelength (although half-wavelength versions can also be built at some cost to gain). As in the previous case, the elements are spaced a half wavelength apart and are fed at their bases through 600 ohm parallel line that is twisted at the ends. Rather than breaking one conductor to connect the feed, this antenna connects the BALUN transformer across the antenna phasing harness wires. Note also that a 1:1 BALUN is used here, rather than a 4:1 device as above. This antenna is capable of as much as 12 dB gain.

THE LAZY-H ARRAY

The lazy-H array is a special case of the doublet array in which the doublets making up the array are one wavelength long (Figure 8.8A), fed in-phase

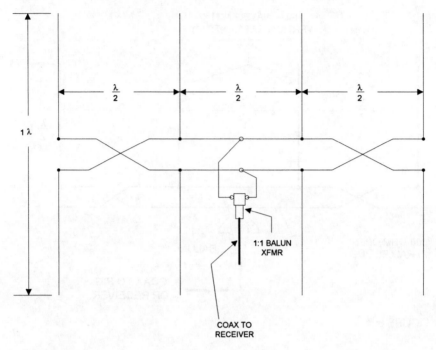

FIGURE 8.7

with each other. The directivity of this antenna is broadside, and so radiates in two directions into and out of the page. This view is as seen from the front or back, so note that the two one-wavelength doublets are stacked one on top of the other.

Gains for the lazy-H antenna vary with the spacing between the doublets (S), and will be as low as 4 dB at $3\lambda/8$ to almost 7 dB at $3\lambda/4$ spacing. The antenna should be installed at least a quarter wavelength above the Earth's surface.

The two doublets in the lazy-H must be fed in-phase with each other. When the feed arrangement is as shown in Figure 8.8A, i.e. with the feedpoint in the center, the transmission line phasing harness is connected straight across (no twisting). If only single-band operation is anticipated, then the feedpoint impedance will be a reasonable match to 75 ohm coaxial cable; a 1:1 BALUN transformer is used to connect the coaxial cable to the phasing harness. For multiband operation, however, it is necessary to use tuned feeders to an antenna tuning unit.

An alternative feed scheme is shown in Figure 8.8B. In this arrangement, the feedpoint is at the feedpoint of one of the doublets, rather than in the center of the harness. In order to preserve the in-phase relationship, the

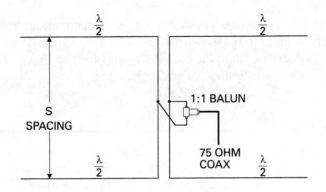

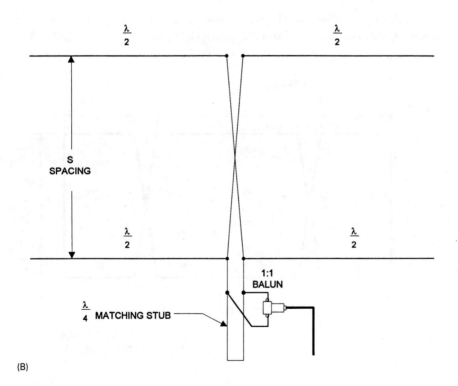

FIGURE 8.8

harness must be twisted once as shown. The feedpoint impedance is around 3000 ohms, so a quarter-wavelength matching stub is needed in order to effect a reasonable match to the coaxial transmission line. Again, for multi-band operation tuned feeders are advisable.

CURTAIN ARRAYS

The curtain arrays in this section are variations of the broadside arrays. The antenna shown in Figure 8.9 is a small **Sterba curtain array**, larger versions of which are often used by high-power international short-wave broadcasters. It can be fed with either 300 ohm twin-lead, or with 75 ohm coaxial cable if a 4:1 BALUN transformer is provided at the feedpoint.

The Sterba curtain antenna can be built of wire or aluminum tubing, although the wire option is probably the most popular at HF, and tubing at VHF. An advantage of these antennas is that they can be built for frequencies in the 6–7 MHz range (where wire construction is preferred), if you have enough room, and also well into the VHF region (in which case aluminum tubing construction is preferred).

The signal is bidirectional, and is broadside to the array (in and out of the page). Gain is of the order of 6 dB, although it rises with added numbers of sections.

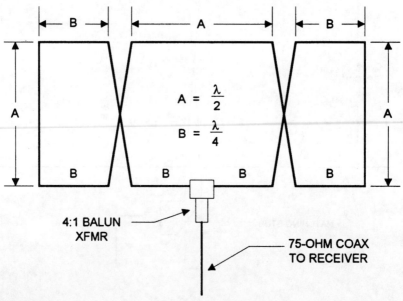

FIGURE 8.9

The antenna in Figure 8.10 (our micro-Sterba) uses elements of two different lengths ($\lambda/2$ and $\lambda/4$), labelled A and B. These lengths can be calculated from

$$A = \frac{149}{F_{MHz}} \text{ meters}$$

$$B = \frac{A}{2}$$

At 16 MHz, these lengths work out to be $A = 9.31$ m, $B = 4.66$ m; at 162 MHz, $A = 0.92$ m and $B = 0.46$ m.

The horizontal distance between horizontal elements should be 10.2–15 cm. In wire antennas, an ordinary end insulator placed between two elements will usually suffice.

The Sterba curtain concept can be extended by adding sections, which of course also increases gain. The version shown in Figure 8.10 provides gain up to 8 dB, and is basically a pair of the previous antennas back to back. The feedpoint is matched using a quarter wavelength stub and a 1:1 BALUN transformer. As with the other antennas, this is a broadside array.

The multisection Sterba curtain array shown in Figure 8.11 uses five loops, and is fed at the end with 300 ohm twin-lead (or a 4:1 BALUN connected to 75 ohm coaxial cable). Gain will be up to 9 dB or so, but at the cost of a lot of space (each A-section is a half wavelength long).

The **Bruce array** is shown in Figure 8.12. This type of array is built using a long wire folded and fed in such a manner that the current nodes (i.e. points of high current) are in the centers of the vertical elements, while the

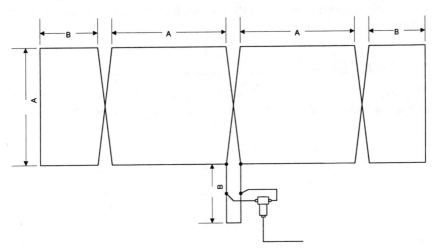

FIGURE 8.10

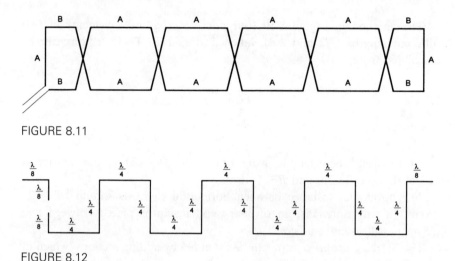

FIGURE 8.11

FIGURE 8.12

voltage nodes (points of low current) are in the centers of the horizontal elements. About 9 dB can be achieved with this antenna.

A variant Bruce array is shown in Figure 8.13. This antenna consists of two cross-connected sections, each of which are made of quarter-wavelength elements. The feedline is connected at the points marked X1–X2. The impedance at this point is around 700–800 ohms. If a 9:1 BALUN transformer is used, then this point can be fed with 75 ohm coaxial cable with only a small voltage standing wave ratio (VSWR), assuming single-band operation. It is also possible to feed this antenna on a single band with a quarter-wavelength matching stub, along with a 1:1 BALUN transformer. If you want to operate on two harmonically related bands, however, tuned feeders and an antenna tuner are required.

'SIX-SHOOTER' ARRAY

The array shown in Figure 8.14A is sometimes called the 'six-shooter' because it consists of six half-wavelength elements. In structure it is much

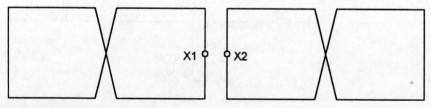

FIGURE 8.13

166 ANTENNA TOOLKIT

like the Sterba curtain shown earlier, with the end elements open circuited rather than connected together. This antenna is capable of gains between 7 and 8 dB.

The lengths of the elements in Figure 8.14A are calculated as shown in the figure. The horizontal elements are each calculated from

$$L_{\text{meters}} = \frac{145}{F_{\text{MHz}}} \text{ meters}$$

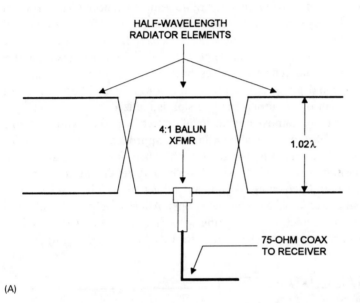

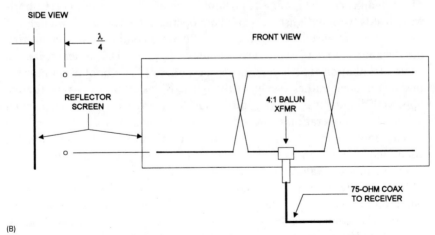

FIGURE 8.14

While the vertical separation between the two rows of horizontal elements is calculated from

$$L_{\text{meters}} = \frac{296}{F_{\text{MHz}}} \text{ meters}$$

These antennas are relatively easy to construct of either wire or tubing, and should be considered whenever you want gain on the cheap.

Like other broadside arrays, the 'six-shooter' is bidirectional, and sends or receives signals in and out of the page as you view the drawing. One variant that I saw on a 10 m amateur radio band 'six-shooter' is shown in Figure 8.14B. In this version, a screen is placed a quarter wavelength behind the antenna. Signals traveling toward the screen from the antenna are reflected back toward the antenna element. Because the spacing is a quarter wavelength, the signal reflected arrives back at the antenna in phase, so reinforces the signal going in the other direction. A gain improvement of 3 dB over the gain of the unscreened version is realized.

The screen can be almost any form of reflective plane. Although one could use sheets of copper or aluminum, this approach is not the best for most purposes. The reason is that the reflector then has a tremendous 'sail area' and so will be susceptible to wind problems. A better solution might be to use metal window screening (if you can still get it), or 'chicken wire' (i.e. the kind of wire used to screen chicken coops). Alternatively, one could also create an effective screen by connecting wires in a square matrix, provided that the gaps are only a small fraction of a wavelength (i.e. $\ll \lambda/8$).

THE BOBTAIL CURTAIN ARRAY

The bobtail curtain (Figure 8.15) is something of a favorite in the HF short-wave bands (with both listeners and ham operators) when very low angles of radiation are required. A low angle is the desired condition for maximizing the distance received by any given antenna (a good 'DX' antenna has an angle of radiation that barely kisses the horizon). The bobtail curtain fits this requirement rather nicely. Although it is easiest to construct for the upper HF bands, it is not beyond reason for many sites for 75/80 m band amateur radio operation or 60 m band short-wave activity.

The bobtail curtain consists of three quarter-wavelength radiators (B) connected at their tops by half-wavelength sections (A). Because the antennas are a quarter wavelength, the proper lengths are of the order of 19.82–21.34 m for the 75/80 m band and 14.63 m for the 60 m band. For the 13 m band, on the other hand, the lengths of the vertical radiators are only 2.9–3.05 m. The connecting half-wavelength sections are twice these lengths. Note that the current which is injected into the center section must split at the top, and only one-half of the line current of the center radiator flows

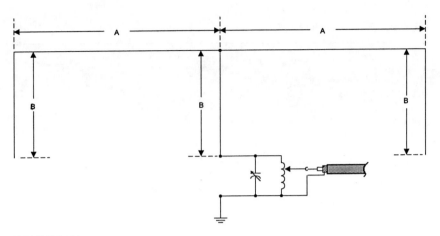

FIGURE 8.15

in each side radiator. This results in a **binomial current distribution** between the elements.

The gain of the bobtail curtain is 7–10 dB, with a figure '8' pattern broadside to the antenna. In trials some years ago, I witnessed a well-made bobtail curtain that only exhibited 5–6 dB (keeping in mind the difficulty of measuring antenna gain *in situ*), but even 5 dB is relatively decent compared to a dipole (nearly an *S*-unit on conservatively specified receivers). The directivity of the bobtail curtain is better than the dipole, but not quite as good as a set of three vertical radiators fed in-phase with each other (which will produce nearly the same pattern).

The bobtail curtain is fed at the base of the center element through a parallel tuned LC tank circuit used for impedance matching. The coaxial cable is connected to either a tap on the inductor (as shown) or to a link or primary winding (a couple of turns of wire) to the main inductor. The values of the inductor and capacitor are set such that the reactances are equal to 1100 ohms. For operation up to 20 MHz, a 50 pF variable capacitor will normally be satisfactory, although at higher frequencies the minimum capacitance of the typical 50 pF unit may be too high for the required capacitance (only a few picofarads). The inductor will vary from about 10 to 50 µH. The various inductances can be provided by either a variable inductor, or a fixed inductor with taps at the required inductance points.

THE THORNE ARRAY BOBTAIL CURTAIN ANTENNA

The Thorne Array Bobtail Curtain (TABC), shown in Figure 8.16, is basically a standard bobtail curtain array turned upside down, and fed

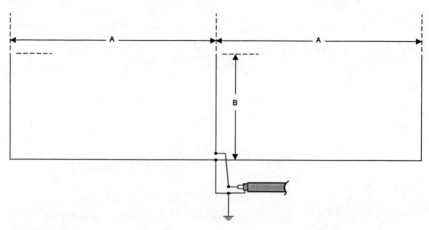

FIGURE 8.16

in a different manner. Bidirectional gains up to 5 dB have been measured on the 15 m band, although I suspect that the gain is due largely to compression of the elevation lobe of the antenna. This antenna has a very low angle of radiation, so works well for long-DX in the upper end of the short-wave spectrum. On the 15 m ham band, from a location in Texas, I have worked very loud Australian and New Zealand stations that were a lot less audible on the quarter-wavelength vertical and dipole antennas at the same location (and received much better signal reports of my own signal).

The TABC consists of three vertical quarter-wavelength ($\lambda/4$) radiators (B) separated by half-wavelength ($\lambda/2$) phasing elements (A). The two outer vertical elements are electrically connected to the corresponding phasing element, but insulated from the center vertical element. The two phasing elements are electrically connected to each other by a very short jumper wire. The center conductor of the 52 ohm coaxial cable to the receiver is connected to the base of the center vertical element, while the coaxial shield is connected to the jumper between phasing elements, and to ground.

The lengths of the two sections are $A = 91/F_{MHz}$ meters and $B = 46/F_{MHz}$ meters. For the 24 MHz band, the horizontal or phasing sections are 3.64 meters long while the vertical sections are 1.84 meters long.

Mechanically, the antenna is basically supported by dipole-like antenna end insulators and ropes to various masts or support structures (trees, buildings, etc.). If you prefer, the vertical sections can be replaced with 12–50 mm aluminum tubing, but you will have to provide some form of mounting for the tubing.

FOLDED DIPOLE 'X-ARRAY'

The array shown in Figure 8.17 uses four folded dipole antennas arranged in an 'X-array' pattern to provide about 6 dB gain. The antenna elements are made from 300 ohm twin-lead. The ends of the twin-lead conductors are shorted together in this type of antenna. The arrangement is two elements in each of two bays. In each bay, the inner ends of the folded dipole elements are spaced 0.2λ apart, while the two bays are spaced 1.25λ apart. The transmission line harnesses (P) are 0.45λ long. These transmission line harnesses are connected in phase with each other at the high impedance side of a 4:1 BALUN transformer; 75 ohm coaxial cable carries the signal to the receiver. All four elements in Figure 8.17 are in the same plane, and the direction of radiation/reception is perpendicular to that plane (i.e. in/out of the page).

PHASED VERTICALS

Figure 8.18 shows the basic two-element phased vertical array. This antenna uses two quarter-wavelength wire radiators erected in a vertical manner, spaced a half wavelength apart. Ropes or heavy twine can be used to support the antenna elements. The antenna element lengths are found from $L_{meters} = 75/F_{MHz}$, while the spacing between them is $D_{meters} = 150/F_{MHz}$.

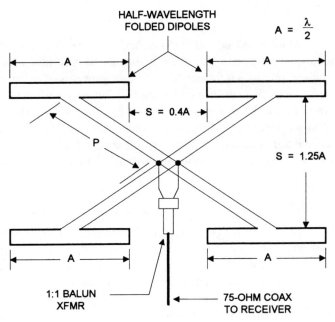

FIGURE 8.17

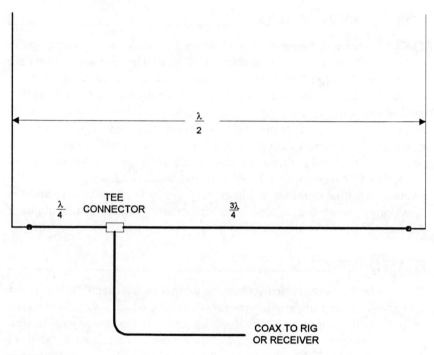

FIGURE 8.18

There are two ways to feed this antenna, and both use a coaxial cable 'tee' connector to split the signal from the rig between the two antennas. If you want to feed the antennas in phase, then the two segments of cable from the 'tee' to the antenna elements are equal lengths. But if you want to feed the elements 180° out of phase, then make one length $\lambda/4$ and the other $3\lambda/4$. In the case of in-phase feed, the directivity and gain are in and out of the page, while when the antennas are in phase the directivity and gain are bidirectional on a line between the two antennas (see again Figures 8.1E and 8.1F).

W8JK ARRAY

The W8JK array (Figure 8.19) is sometimes called the 'flat-top' beam. It is a bidirectional, end-fire array that consists of two half-wavelength in-phase collinear doublets, in parallel and fed 180° out of phase with each other, spaced between $\lambda/8$ and $\lambda/4$ apart. Each element in each collinear doublet is $\lambda/4$ long (dimension A). The view in Figure 8.20 is from above, so the plane of the dipoles is horizontal to the Earth's surface. The gain will be from 5.7 dBi at $\lambda/4$ spacing to 6.2 dBi at $\lambda/8$ spacing. Note that the spacing

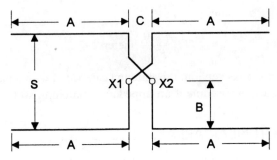

FIGURE 8.19

between the elements of the same dipole is of the order of

$$C = \frac{8.5}{F_{MHz}} \text{ meters}$$

The length of each wire element in the dipoles is found from:

$$A = \frac{71.5}{F_{MHz}} \text{ meters}$$

The feedline phasing harness between the center-feed points on the two dipoles is made of 450–600 ohm parallel transmission line (or 450 ohm twin-lead). The parallel feedline to the transmitter or receiver is attached to the phasing harness at points X1–X2, at a distance (B) of about $S/2$.

The impedance at the feedpoint (X1–X2) is several thousand ohms, so will necessarily create a rather high VSWR. An impedance matching stub can be provided at the feedpoint to improve the match to the transmission line. However, if tuned feeders are used then this antenna is capable of multiband (harmonically related) operation.

An end-fed version of the W8JK array is shown in Figure 8.20. In this case, the C-dimension is the same as calculated above, but the A-dimension

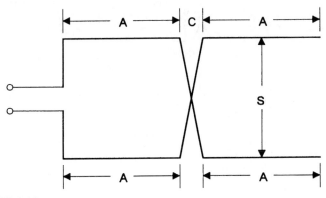

FIGURE 8.20

is not quite twice as long:

$$A = \frac{110}{F_{\text{MHz}}} \text{ meters}$$

As with the previous version, this antenna can be fed with a parallel tuned feeder, or with a matched line if an impedance matching stub is provided.

STACKED COLLINEAR ARRAYS

The collinear array antenna can be stacked as in Figure 8.21. This antenna consists of two collinear arrays (similar to Figure 8.5). The lengths of A, B and C are found from:

$$A = \frac{144}{F_{\text{MHz}}}$$

$$B = \frac{73}{F_{\text{MHz}}}$$

$$C = \frac{150}{F_{\text{MHz}}}$$

where A, B and C are in meters, and F_{MHz} is frequency in megahertz.

The two antennas are shown stacked vertically, although they can be stacked horizontally as well (although with a different pattern).

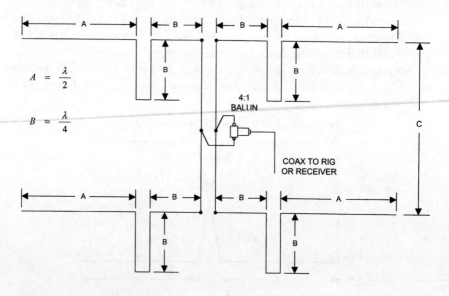

FIGURE 8.21

LARGE REFLECTOR ARRAY ANTENNA

Figure 8.22 shows a large-scale reflector array. It consists of a group of sixteen half-wavelength elements stacked vertically and horizontally. The elements are cross-connected with phasing harnesses. The spacing between elements is half wavelength, although the formula used to calculate this half wavelength is different from the half wavelength of the wire elements:

Wire elements:

$$\frac{\lambda}{2} = \frac{143}{F_{\text{MHz}}}$$

Spacing:

$$\frac{\lambda}{2} = \frac{150}{F_{\text{MHz}}}$$

The reflector screen can be made of chicken wire or any other screening material in which the holes are not more than about $\lambda/12$ in size. It is sized at least $\lambda/8$ larger than the outside dimensions of the array itself. The reflector screen is located a quarter wavelength behind the antenna array elements. The antenna array elements are made of either thin (#10 to #18) or thick wire (#000 to #10).

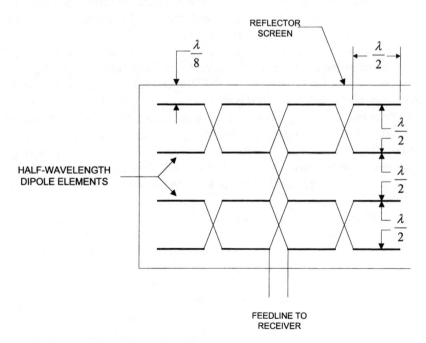

FIGURE 8.22

CHAPTER 9

Small loop antennas

In this chapter you will learn about small loop receiving antennas. The small loop antenna is almost ideal for DXing the crowded AM broadcast band and low frequency 'tropical' bands. These antennas are fundamentally different from the large loop types previously discussed and are very often the antenna of choice for low frequency work.

Large loop antennas are 0.5λ or larger and respond to the electrical field component of the electromagnetic wave. Small loop antennas, on the other hand, are $<0.1\lambda$ (some sources say 0.17λ and $<0.22\lambda$) and respond to the magnetic field component of the electromagnetic wave. One principal difference between the large loop and the small loop is found when examining the radio frequency current induced in the loop when a radio signal intersects it. In a large loop, the dimensions in each section are an appreciable portion of one wavelength, so the current will vary from one point in the conductor to another. But in a small loop, the current is the same throughout the entire loop.

The differences between small loops and large loops show up in some interesting ways, but perhaps the most striking are the directions of maximum response – the main lobes – and the directions of the nulls. Both types of loop produce figure-of-eight patterns, but in directions at right angles with respect to each other. The large loop antenna produces main lobes *orthogonal*, at right angles or 'broadside' to, the plane of the loop. Nulls are off the sides of the loop. The small loop, however, is exactly the opposite: the main lobes are off the sides of the loop (in the direction of the loop plane), and the nulls are broadside to the loop plane (see Figure 9.1A).

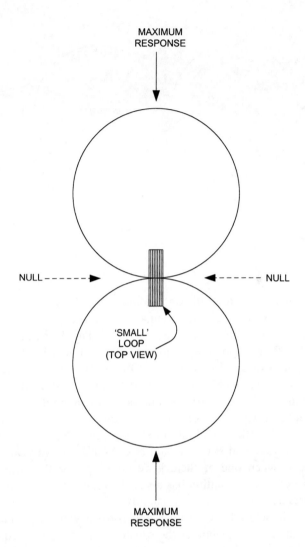

FIGURE 9.1A

Don't confuse small loop behavior with the behavior of the loopstick antenna. Loopstick antennas are made of coils of wire wound on a ferrite or powdered iron rod. The direction of maximum response for the loopstick antenna is broadside to the rod with deep nulls off the ends (Figure 9.1B). Both loopsticks and small wire loops are used for radio direction-finding and for shortwave, low frequency medium wave, AM broadcast band, and VLF DXing.

The nulls of a loop antenna are very sharp and very deep. Small changes of pointing direction can make a profound difference in the response of the antenna. If you point a loop antenna so that its null is aimed at a strong

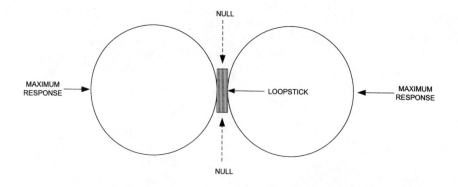

FIGURE 9.1B

station, the signal strength of the station appears to drop dramatically at the center of the notch. But turn the antenna only a few degrees one way or the other, and the signal strength increases sharply. The depth of the null can reach 10 to 15 dB on sloppy loops and 30 to 40 dB on well-built loops (20 dB is a very common value). I've seen claims of 60 dB nulls for some commercially available loop antennas. The construction and uniformity of the loop are primary factors in the sharpness and depth of the null.

One of the characteristics of the low frequency bands in which small loops are effective is the possibility of strong local interference smothering weaker ground wave and sky wave stations. As a result, you can't hear co-channel signals when one of them is very strong and the other is weak. Similarly, if a co-channel station has a signal strength that is an appreciable fraction of the desired signal, and is slightly different in frequency, then the two signals will heterodyne together and form a whistling sound in the receiver output. The frequency of the whistle is an audio tone equal to the difference in frequency between the two signals. This is often the case when trying to hear foreign broadcast band signals on frequencies (called *split frequencies*) between the standard 10 kHz spacing used in North and South America. The directional characteristics of the loop can help if the loop null is placed in the direction of the undesired signal.

Loops are used mainly in the low frequency bands even though such loops are either physically larger than high frequency loops or require more turns of wire. Loops have been used as high as VHF and are commonly used in the 10 meter ham band for such activities as hidden transmitter hunts.

Let's examine the basic theory of small loop antennas, and then take a look at some practical construction methods.

AIR CORE FRAME LOOPS ('BOX' LOOPS)

A wire loop antenna is made by winding a large coil of wire, consisting of one or more turns, on some sort of frame. The shape of the loop can be circular, square, triangular, hexagonal or octagonal. For practical reasons, the square loop seems to be most popular. With one exception, the loops considered in this section will be square so you can easily duplicate them.

The basic form of the simplest loop is shown in Figure 9.2. This loop is square, with sides the same length 'A' all around. The width of the loop ('B') is the distance from the first turn to the last turn in the loop, or the diameter of the wire if only one turn is used. The turns of the loop in Figure 9.2 are *depth wound,* meaning each turn of the loop is spaced in a slightly different parallel plane. The turns are spaced evenly across distance 'B'. Alternatively, the loop can be wound such that the turns are in the same plane (this is called *planar winding).* In either case, the sides of the loop ('A') should be not less than five times the width ('B'). There seems to be little difference between depth and planar wound loops. The far-field patterns of the different shape loops are nearly the same if the respective cross sectional areas (πr^2 for circular loops and A^2 for square loops) are $<\lambda^2/100$.

The actual voltage across the output terminals of an untuned loop is a function of the angle of arrival of the signal a, as well as the strength of the signal and the design of the loop. The voltage V_o is given by:

$$V_o = \frac{2\pi A N E_f \cos(\alpha)}{\lambda}$$

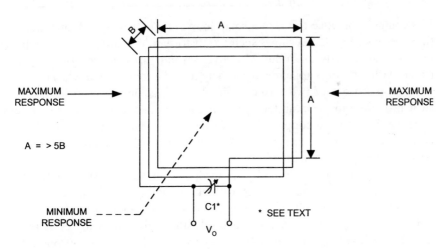

FIGURE 9.2

where:
- V_o is the output voltage of the loop
- A is the area of the loop in square meters (m²)
- N is the number of turns of wire in the loop
- E_f is the strength of the signal in volts per meter (V/m)
- a is the angle of arrival of the signal
- λ is the wavelength of the arriving signal

Loops are sometimes specified in terms of the *effective height* of the antenna. This number is a theoretical construct that compares the output voltage of a small loop with a vertical piece of the same kind of wire that has a height of:

$$H_{\text{eff}} = \frac{2\pi NA}{\lambda}$$

If a capacitor (such as C1 in Figure 9.2) is used to tune the loop, then the output voltage V_o will rise substantially. The output voltage found using the first equation is multiplied by the loaded Q of the tuned circuit, which can be from 10 to 100 (if the antenna is well constructed):

$$V_o = \frac{2\pi ANE_f \, Q \cos(\alpha)}{\lambda}$$

Even though the output signal voltage of tuned loops is higher than that of untuned loops, it is nonetheless low compared with other forms of antenna. As a result, a loop preamplifier is usually needed for best performance.

TRANSFORMER LOOPS

It is common practice to make a small loop antenna with two loops rather than just one. Figure 9.3 shows such a transformer loop antenna. The main loop is built exactly as discussed above: several turns of wire on a large frame, with a tuning capacitor to resonate it to the frequency of choice. The other loop is a one or two turn coupling loop. This loop is installed in very close proximity to the main loop, usually (but not necessarily) on the inside edge not more than a couple of centimeters away. The purpose of this loop is to couple signal induced from the main loop to the receiver at a more reasonable impedance match.

The coupling loop is usually untuned, but in some designs a tuning capacitor (C2) is placed in series with the coupling loop. Because there are many fewer turns on the coupling loop than the main loop, its inductance is considerably smaller. As a result, the capacitance to resonate is usually much larger. In several loop antennas constructed for purposes of researching this chapter, I found that a 15-turn main loop resonated in the

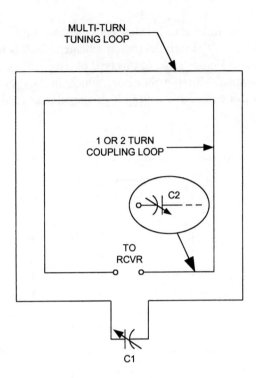

FIGURE 9.3

AM broadcast band with a standard 365 pF capacitor, but the two turn coupling loop required three sections of a ganged 3 × 365 pF capacitor connected in parallel to resonate at the same frequencies.

In several experiments, I used computer ribbon cable to make the loop turns. That type of cable consists of anywhere from eight to 64 parallel insulated conductors arranged in a flat ribbon shape. Properly interconnected (of which more later), the conductors of the ribbon cable form a continuous loop. It is no problem to take the outermost one or two conductors on one side of the wire array and use it for a coupling loop.

TUNING SCHEMES FOR LOOP ANTENNAS

Loop performance is greatly enhanced by tuning the inductance of the loop to the desired frequency. The bandwidth of the loop is reduced, which reduces front-end overload. Tuning also increases the signal level available to the receiver by a factor of 10 to 100 times. Although tuning can be a bother if the loop is installed remotely from the receiver, the benefits are well worth it in most cases.

There are several different schemes available for tuning, and these are detailed in Figure 9.4. The parallel tuning scheme, which is by far the most popular, is shown in Figure 9.4A. In this type of circuit, the capacitor (C1) is connected in parallel with the inductor, which in this case is the loop. Parallel resonant circuits have a very high impedance to signals on their

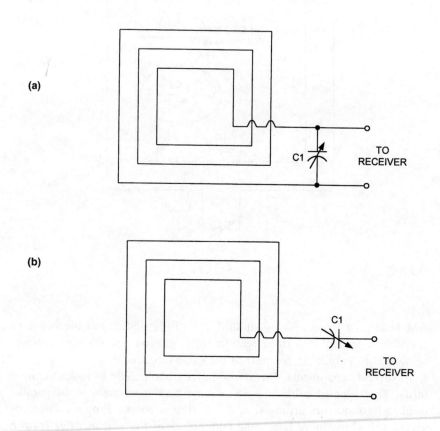

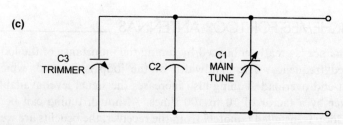

FIGURE 9.4

182 ANTENNA TOOLKIT

Jo Coleman
Information Update Service
Butterworth-Heinemann
FREEPOST SCE 5435
Oxford
Oxon
OX2 8BR
UK

FOR CARDS OUTSIDE THE UK, PLEASE AFFIX A POSTAGE STAMP

Keep up-to-date with the latest books in your field.

Visit our website and register now for our FREE e-mail update service, or join our mailing list and enter our monthly prize draw to win £100 worth of books. Just complete the form below and return it to us now! (FREEPOST if you are based in the UK)

www.bh.com

Please Complete In Block Capitals

Title of book you have purchased:..

....

Subject area of interest:...

Name:..

Job title:..

Business sector (if relevant):..

Street:...

Town:... County:..

Country:... Postcode:..

Email:..

Telephone:...

How would you prefer to be contacted: Post ☐   e-mail ☐   Both ☐

Signature:... Date:...

☐ Please arrange for me to be kept informed of other books and information services on this and related subjects (✔ box if not required). This information is being collected on behalf of Reed Elsevier plc group and may be used to supply information about products by companies within the group.

FOR OFFICE USE ONLY

Butterworth-Heinemann,
a division of Reed Educational
& Professional Publishing Limited.
Registered office: 25 Victoria Street,
London SW1H 0EX.
Registered in England 3099304.
VAT number GB: 663 3472 30.

resonant frequency, and a very low impedance to other frequencies. As a result, the voltage level of resonant signals is very much larger than the voltage level of off-frequency signals.

The series resonant scheme is shown in Figure 9.4B. In this circuit, the loop is connected in series with the capacitor. A property of series resonant circuits is that it offers a high impedance to all frequencies except the resonant frequency (exactly the opposite of the case of parallel resonant circuits). As a result, current from the signal will pass through the series resonant circuit at the resonant frequency, but off-frequency signals are blocked by the high impedance.

There is a wide margin for error in the inductance of loop antennas, and even the precise-looking equations to determine the required values of capacitance and inductance for proper tuning are actually only estimations. The exact geometry of the loop 'as built' determines the actual inductance in each particular case. As a result, it is often the case that the tuning provided by the capacitor is not as exact as desired, so some form of compensation is needed. In some cases, the capacitance required for resonance is not easily available in a standard variable capacitor and some means must be provided for changing the capacitance. Figure 9.4C shows how this is done. The main tuning capacitor can be connected in either series or parallel with other capacitors to change the value. If the capacitors are connected in parallel, then the total capacitance is increased (all capacitances are added together). But if the extra capacitor is connected in series then the total capacitance is reduced. The extra capacitors can be switched in and out of a circuit to change frequency bands.

Tuning of a remote loop can be a bother if done by hand, so some means must be found to do it from the receiver location (unless you enjoy climbing into the attic or onto the roof). Traditional means of tuning called for using a low speed DC motor, or stepper motor, to turn the tuning capacitor. A very popular combination was the little 1 to 12 RPM motors used to drive rotating displays in retail store show windows. But that approach is not really needed today. We can use varactor voltage variable capacitance diodes to tune the circuit.

A varactor works because the junction capacitance of the diode is a function of the applied reverse bias voltage. A high voltage (such as 30 volts) drops the capacitance while a low voltage increases it. Varactors are available with maximum capacitances of 22, 33, 60, 100, and 400 pF. The latter are of most interest to us because they have the same range as the tuning capacitors normally used with loops. Look for service shop replacement diodes intended for use in AM broadcast band radios. A good selection, which I have used, is the NTE-618 device. It produces a high capacitance >400 pF, and a low of only a few picofarads over a range of 0 to 15 volts.

Figure 9.5 shows how a remote tuning scheme can work with loop antennas. The tuning capacitor is a combination of a varactor diode and two optional capacitors: a fixed capacitor (C1) and a trimmer (C2). The DC tuning voltage (V_t is provided from the receiver end from a fixed DC power supply ($V+$). A potentiometer (R1) is used to set the voltage to the varactor, hence also to tune the loop. A DC blocking capacitor (C3) keeps the DC tuning voltage from being shorted out by the receiver input circuitry.

THE SQUARE HOBBY BOARD LOOP

A very common way to build a square loop antenna is to take two pieces of thin lumber, place them in a cross shape, and then wind the wire around the ends of the wooden arms. This type of antenna is shown in Figure 9.6. The

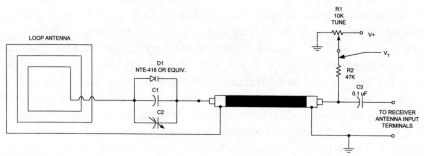

FIGURE 9.5

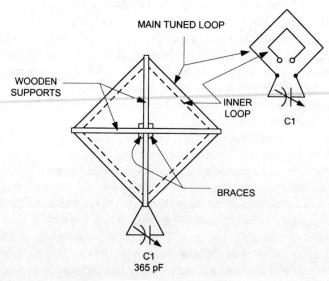

FIGURE 9.6

184 ANTENNA TOOLKIT

wooden supports can be made of 1″ × 2″ lumber, or some other stock. A test loop made while researching this book was made with two 3/16″ × 3″ × 24″ Bass wood 'hobby board' stock acquired from a local hobby shop. Model builders use this wood as a stronger alternative to balsa wood.

The electrical circuit of the hobby board loop shown in Figure 9.6 is a transformer loop design. The main tuned loop is on the outside and consists of ten turns of #26 enameled wire. It is tuned by a 365 pF capacitor. The inner loop is used for coupling to the receiver and consists of a single turn of #22 insulated solid hook-up wire.

Details for the boards are shown in Figure 9.7. Each board is 24 inches (61 cm) long. At the mid-point (12″ or 30.5 cm), there is a $\frac{1}{4}$ inch (0.635 cm) wide, 1.5 inch (3.8 cm) long slot cut. These slots are used to mate the two boards together.

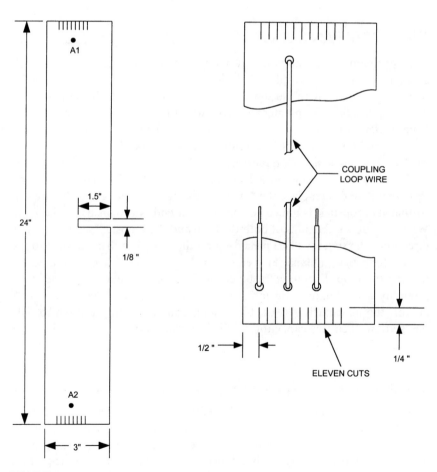

FIGURE 9.7

At each end there are ten tiny slits made by a jeweler's saw (also called a 'jig saw' in hobbyist circles) with a thin blade. These slits are just wide enough to allow a single #26 wire to be inserted without slipping. The slits are about $\frac{1}{4}$ inch (0.635 cm) long, and are $\frac{1}{4}$ inch (0.635 cm) apart. There are ten slits on both ends of the horizontal piece while the vertical piece has ten slits on the top end and 11 slits on the bottom end. The reason for offsetting the wire slits is to allow room on the other side of the 3 inch width of the vertical member for a mounting stick.

When assembling the antenna, use wood glue on the mating surfaces, square them to be at right angles to each other, and clamp the two pieces in a vice or with C-clamps for 30 minutes (or longer if the glue maker specifies). Next, glue the support blocks into place and clamp them for a similar period.

THE SPORTS FAN'S LOOP

OK, sports fans, what do you do when the best game of the week is broadcast only on a low-powered AM station – and you live at the outer edge of their service area where the signal strength leaves much to be desired? You use the sports fan's loop antenna, that's what! I first learned of this antenna from a friend of mine, a professional broadcast engineer, who worked at a religious radio station that had a pipsqueak signal but lots of fans. It really works – *one might say it's a miracle*.

The basic idea is to build a 16-turn, 24 inch (61 cm) square tuned loop (Figure 9.8) and then place the AM portable radio at the center (Figure 9.9) so that its loopstick is aimed so that its null end is broadside of the loop. When you do so, the nulls of both the loop and the loopstick are in the same direction. The signal will be picked up by the loop and then coupled to the radio's loopstick antenna. Sixteen-conductor ribbon cable can be used for making the loop. For an extra touch of class, place the antenna and radio assembly on a dining room table 'Lazy Susan' to make rotation easier. A 365 pF tuning capacitor is used to resonate the loop. If you listen to only one station, then this capacitor can be a trimmer type.

SHIELDED LOOP ANTENNAS

The loop antennas discussed thus far in this chapter have all been unshielded types. Unshielded loops work well under most circumstances, but in some cases their pattern is distorted by interaction with the ground and nearby structures (trees, buildings, etc.). In my own tests, trips to a nearby field proved necessary to measure the depth of the null because of

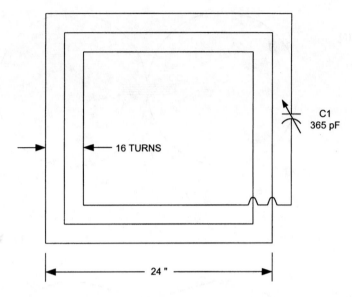

FIGURE 9.8

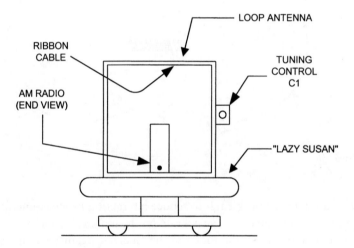

FIGURE 9.9

interaction with the aluminum siding on my house. Figure 9.10 shows two situations. In Figure 9.10A, we see the pattern of the normal 'free-space' loop, i.e., a perfect figure-of-eight pattern. But when the loop interacts with the nearby environment, the pattern distorts. In Figure 9.10B we see some filling of the notch for a moderately distorted pattern. Some interactions are so severe that the pattern is distorted beyond all recognition.

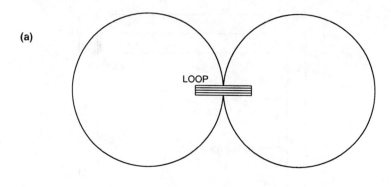

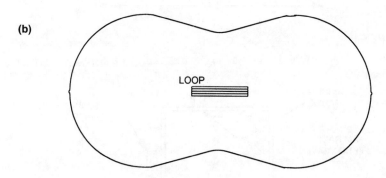

FIGURE 9.10

The solution to the problem is to reduce interaction by shielding the loop, as in Figure 9.11. Loop antennas operate on the magnetic component of the electromagnetic wave, so the loop can be shielded against voltage signals and electrostatic interactions. In order to prevent harming the ability to pick up the magnetic field, a gap is left in the shield at one point.

There are several ways to shield a loop. You can, for example, wrap the loop in adhesive-backed copper foil tape. Alternatively, you can wrap the loop in aluminum foil and hold it together with tape. Another method is to insert the loop inside a copper or aluminum tubing frame. The list seems endless. Perhaps one of the most popular methods is to use coaxial cable to make a large single turn loop. Figure 9.12 shows this type of loop made with RG-8/U or RG-11/U coaxial cable. The cable is normally supported by wooden cross arms, as in the other forms of loop, but they are not shown

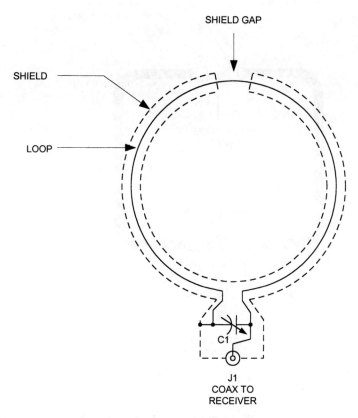

FIGURE 9.11

here for sake of simplicity. Note that, at the upper end, the coaxial cable shields are not connected.

Another example is the antenna shown in Figure 9.13. This antenna is made of wide metal conductors. Examples include the same type of hobbyist's brass stock as used above. It can also be copper foil or some other stock that can be soldered. Some electronic parts stores sell adhesive backed foil stock used for making printed circuit boards. The foil can be glued to some flat insulating surface. Although K-inch plywood springs to mind immediately, another alternative is found in artists' supplies stores. Ordinary poster board is too floppy to stand up, but poster board glued to a Styrofoam backing can be used. It is extremely easy to work with using X-acto knives and other common household tools.

Two controls are used on this antenna. Capacitor C1 tunes the loop to the resonant frequency of the desired station. Potentiometer R1 is used as a phasing control. The dimensions of the antenna are not terribly critical,

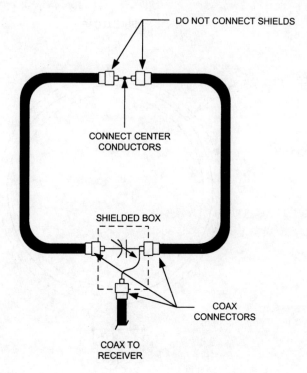

FIGURE 9.12

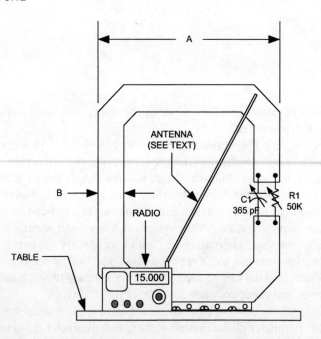

FIGURE 9.13

although some guidelines are in order. In the Villard article, he recommended a 40 cm (15.75-inch) square loop ('A'). If the loop is 7.62 cm (three inches) wide, the antenna will resonate at 15 MHz with around 33 pF of capacitance. If the dimensions are increased to A = 91 cm (36 inches) and B = 10.16 cm (four inches), then the inductance increases and only 28 pF are needed at 15 MHz. The larger size loop can be used at lower frequencies as well. For example, the 91 cm loop will resonate at 6 MHz with 177 pF.

To use this antenna, position the radio's telescopic antenna close and adjacent to the loop *but not touching it*. The loop antenna can be rotated to find the best position to either null or enhance a particular station. The 'Lazy Susan' idea will work well in this case.

TESTING YOUR LOOP ANTENNA

When each loop prototype was completed, I tested it on the AM broadcast band over several evenings. The same procedure can be used with any loop. A strong local signal at 1310 kHz served to check the pattern. The station and my home were located on US Geological Survey 7.5 minute quadrangle maps of my area (or the equivalent Ordnance Survey maps in the UK). The maps had adjacent coverage, so the compass bearing from my location to the station could be determined. Checking the antenna showed an S7/S8 signal when the loop was endwise to the station – that is, the station was in one of its lobes. Rotating the loop so that its broadside faced the direction of the station dropped the signal strength to less than S1, and frequently bottomed out the meter. Because my receiver has a 3 dB/S-unit calibration on the S-meter, I figured the null to be more than 20 dB, although it will take a bit more experimentation to find the actual depth. This test is best done during daylight hours, I found out, because there is always a residual sky wave cacophony on the AM band that raises the S-meter 'floor' in the null.

USING A LOOP ANTENNA

Most readers will use a loop for DXing rather than hidden transmitter hunting, navigation, or other RDF purposes. For the DXer, there are actually two uses for the loop. One is when you are a renter or live in a community that has routine covenants against outdoor antennas. In this situation, the loop will serve as an active antenna for receiving AM broadcast band and other low frequency signals without the neighbors or landlord becoming annoyed.

The other use is illustrated by the case of a friend of mine. He regularly tunes in clear channel WSM (650 kHz, Nashville) in the wee hours between

Saturday evening ('Grand Ole Opry' time) and dawn. However, that 'clear' channel of WSM isn't really so clear, especially without a narrow filter in the receiver. He uses a loop antenna to null out a nearby 630 kHz signal that made listening a bit dicey, and can now tape his 1940s/1950s vintage country music.

It isn't necessary to place the desired station directly in the main lobes off the ends of the antenna, but rather place the nulls (broadside) in the direction of the offending station that you want to eliminate. So what happens if the offending station and the desired station are in a direct line with each other with your receiving location in the middle between them? Both nulls and lobes on a loop antenna are bidirectional, so a null on the offending station will also null the desired station in the opposite direction.

One method is to use a sense antenna to spoil the pattern of the loop to a cardioid shape. Another method is to use a spoiler loop to null the undesired signal. The spoiler loop is a large box loop placed one to three feet (found experimentally) behind the reception loop in the direction of the offending signal. This method is detailed in Figure 9.14. The small loopstick may be the antenna inside the receiver, while the large loop is a box loop such as the sports fan's loop. The large box loop is placed about one to three feet behind the loopstick and in the direction of the offending station. The angle with respect to the line of centers should be 60° to 90°, which is also found experimentally. It's also possible to use two air core loops to produce an asymmetrical receiving pattern.

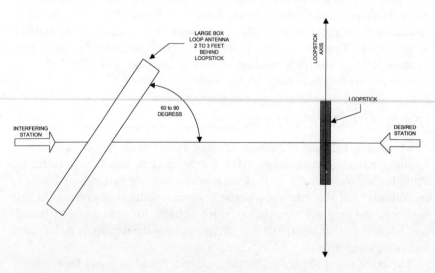

FIGURE 9.14

LOOP PREAMPLIFIERS

All small loop antennas produce a weak output signal, so a loop preamplifier is indicated for all but the most sensitive receivers. The preamplifier can be mounted either at the receiver or the antenna, but it is most effective when mounted at the antenna (unless the coax to the receiver is short).

SHARPENING THE LOOP

Many years ago the *Q-multiplier* was a popular add-on accessory for a communications receiver. These devices were sold as Heathkits and many construction projects were seen in magazines and amateur radio books. The Q-multiplier has the effect of seeming to greatly increase the sensitivity of a receiver, as well as greatly reducing the bandwidth of the front end. Thus, it allows better reception of some stations because of increased sensitivity and narrowed bandwidth.

A Q-multiplier is an active electronic circuit placed at the antenna input of a receiver. It is essentially an Armstrong oscillator, as shown in Figure 9.15, that doesn't quite oscillate. These circuits have a tuned circuit (L1/C1) at the input of an amplifier stage, and a feedback coupling loop (L3). The

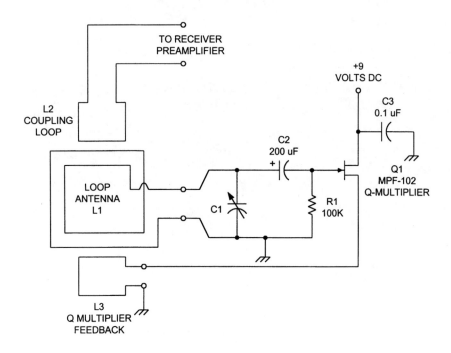

FIGURE 9.15

degree of feedback is controlled by the coupling between L1 and L3. The coupling is varied both by varying how close the two coils are, and their relative orientation with respect to each other. Certain other circuits use a series potentiometer in the L3 side that controls the amount of feedback.

The Q-multiplier is adjusted to the point that the circuit is just on the verge of oscillating, but not quite. As the feedback is backed away from the threshold of oscillation, but not too far, the narrowing of bandwidth occurs as does the increase in sensitivity. It takes some skill to operate a Q-multiplier, but it is easy to use once you get the hang of it and is a terrific accessory for any loop antenna.

CHAPTER 10

Yagi beam antennas

The Yagi beam antenna (more correctly, the Yagi–Uda antenna, after both of the designers of Tohoku University in Japan 1926) is *unidirectional*. It can be vertically polarized or horizontally polarized with little difference in performance (other than the polarization!). The Yagi antenna can be rotated into position with little effort. Yet the Yagi antenna shows power gain (so it puts out and receives a stronger signal), reduces the interfering signals from other directions, and is relatively compact.

COMPOSITION OF A BEAM ANTENNA

The Yagi antenna is characterized by a single driven element which takes power from the transmitter (or is connected to the receiver), plus one or more *parasitic elements*. The parasitic elements are not connected to the driven element, but rather receive their power from the driven element by indirect means. The indirect means is that they intercept the signal, and then re-radiate them.

The minimalist two element beam antenna may be composed of either a driven element and a reflector, or a driven element and a director. The reflector and directors are known as parasitic elements.

The *parasitic reflector* is three to five per cent *longer* than the half wavelength driven element. It provides power gain in the direction away from itself. It is inductive in reactance and lagging in phase.

The *parasitic director* is three to five per cent *shorter* than the half wavelength driven element. It provides power gain in its own direction. It is capacitive in reactance and leading in phase.

The factors that affect the phase difference between the direct and re-radiated signals is determined principally by the *element length* and the *spacing* between the elements. Proper adjustment of these factors determines the gain and the front-to-back ratio that is available.

The presence of a parasitic element in conjunction with a driven element tends to reduce the feedpoint impedance of the driven element for close spacings ($<\lambda/2$) and increase it for greater than $\lambda/2$ spacing. In general, beam antennas have element spacing of 0.1λ to 0.25λ (with 0.15λ to 0.19λ being common), so the impedance will be lower than the nominal impedance for a half wavelength dipole.

TWO-ELEMENT YAGI ARRAY

Figure 10.1 shows the two-element Yagi array antenna. This particular one uses a driven element and a director, so the direction of maximum signal is in the direction of the director. The gain of a two-element Yagi is about 5.5 dBd (gain above a dipole) for spacing less than 0.1λ and the parasitic element is a director. For the case where a reflector is used, the gain peak is 4.7 dBd at about 0.2λ spacing.

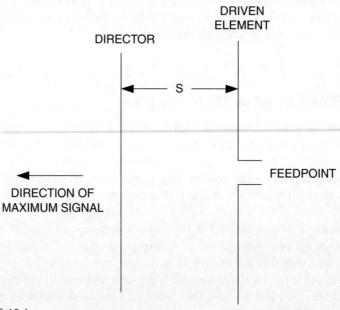

FIGURE 10.1

The difference between reflector and director usage is quite profound. The usual curve shows the director with higher gain, but it is more responsive to element spacing. The reflector has less gain, but is more tolerant of spacing errors.

The front-to-back ratio of the beam antenna is poor for two-element antennas. A compromise spacing of 0.15λ provides front-to-back ratios of 5 to 12 dB.

The feedpoint resistance of the antenna is clearly not 73 ohms as would be implied by the use of a half wavelength dipole for a driven element. The feedpoint impedance will vary roughly linearly from about 5 ohms at a spacing of 0.05λ to about 30 ohms for spacings of about 0.15λ. Above 0.15λ the differences between director and reflector implementations takes place. A reflector two-element beam feedpoint impedance will increase roughly linearly from 30 ohms at 0.15λ to about 45 ohms at 0.25λ spacing. The director implementation is a little less linearly related to spacing, but varies from about 30 ohms at 0.15λ to about 37 ohms at 0.25λ spacing.

Element lengths

The element lengths for a two-element Yagi beam are given below:

Director: $$\text{Director} = \frac{138.6}{F_{MHz}}$$

Driven element: $$\text{D.E.} = \frac{146}{F_{MHz}}$$

Spacing: $$\text{Spacing} = \frac{44.98}{F_{MHz}}$$

Where:
Director is the length of the director
D.E. is the length of the driven element in meters (m)
Spacing is the spacing between the elements in meters (m)
F_{MHz} is the frequency in megahertz.

These element lengths will result in 0.15λ spacing, which is considered about ideal.

THREE-ELEMENT YAGI BEAM

Figure 10.2 shows a Yagi antenna made up of a half wavelength driven element, a reflector and a director. The gain of the array and the front-to-back ratio peaks at a particular boom length (boom not shown), which is indicative of the spacing between the elements. Maximum gain occurs at a boom length of 0.45λ. An example of a three-element Yagi antenna built on a 0.3λ boom will provide 7 to 8 dBd forward gain, and a front-to-back ratio of 15 to 28 dB depending on the element tuning.

The feedpoint impedance of the three-element beam is about 18 to 25 ohms, so some means must be provided for adjusting the impedance to the 52 ohm coaxial cable.

Element lengths

Director: $\quad$ Director $= \dfrac{140.7}{F_{MHz}}$

Driven element: $\quad$ D.E. $= \dfrac{145.7}{F_{MHz}}$

Reflector: $\quad$ Reflector $= \dfrac{150}{F_{MHz}}$

Spacing: $\quad$ Spacing $= \dfrac{43.29}{F_{MHz}}$

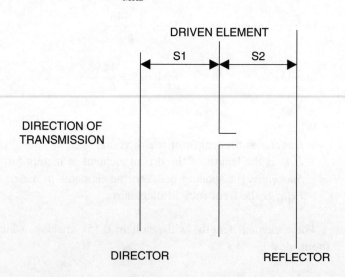

FIGURE 10.2

Where:
   Director is the length of the director in meters (m)
   D.E. is the length of the driven element in meters (m)
   Reflector is the length of the reflector element in meters (m)
   Spacing is the spacing of the elements in meters (m).

FOUR-ELEMENT YAGI ANTENNA

Figure 10.3 shows a four-element Yagi antenna. There is a tremendous increase in forward gain by adding a second director to the three-element case, but the front-to-back ratio is poorer unless the spacing is increased from 0.15λ to about 0.25λ. When all elements are spaced 0.15λ apart, the front-to-back ratio is only about 10 dB, but at 0.25λ the front-to-back ratio increases to 27 dB.

Element lengths

The dimensions calculated from the equations below will yield a forward gain of about 9.1 dBd, with a front-to-back ratio of about 27 dB.

Director: $\quad\quad\quad\quad$ $\text{Director} = \dfrac{138.93}{F_{\text{MHz}}}$

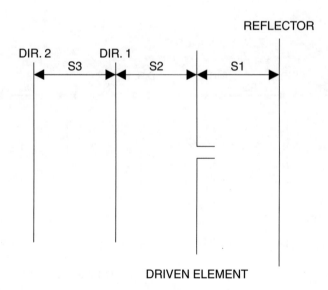

FIGURE 10.3

Driven element: $$\text{D.E.} = \frac{143.65}{F_{\text{MHz}}}$$

Spacing S1: $$S1 = \frac{60.95}{F_{\text{MHz}}}$$

Spacings S2 and S3: $$S2 = S3 = \frac{75}{F_{\text{MHz}}}$$

Where:
 Director is the length of the director element in meters (m)
 D.E. is the length of the driven element in meters (m)
 Reflector is the length of the reflector element in meters (m)
 S1, *S2* and *S3* are in meters (m).

SIX-ELEMENT YAGI ANTENNA

Computer studies of Yagi antenna arrays demonstrate that five-element antennas are little more than four-element antennas, despite the extra director. But the addition of two additional directors adds significantly to the gain. The six-element antenna (Figure 10.4) shows a gain of nearly 10.5 dBd, and a front-to-back ratio of nearly 35 dB. Unfortunately, a six-element antenna is quite large, even with 0.15λ element spacing. At 14 MHz the antenna soars to about 16 meters boom length.

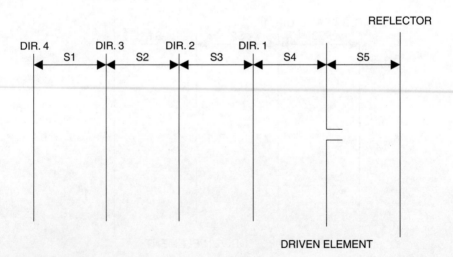

FIGURE 10.4

Element lengths

Director: $\quad$ Director $= \dfrac{134.39}{F_{MHz}}$

Driven element: $\quad$ D.E. $= \dfrac{144.05}{F_{MHz}}$

Reflector: $\quad$ Reflector $= \dfrac{148.56}{F_{MHz}}$

Element spacing: $\quad$ Spacing $= \dfrac{44.8}{F_{MHz}}$

Where:
  Director is the length of the director elements in meters (m)
  D.E. is the length of the driven element in meters (m)
  Reflector is the length of the reflector element in meters (m)
  Spacing is the spacing between the elements in meters (m).

IMPEDANCE MATCHING THE BEAM ANTENNA

The feedpoint impedance of most beam antennas is lower than the feedpoint impedance of a half wavelength dipole (72 ohms), despite the fact that the half wavelength dipole is a driven element. The feedpoint impedance may be as low as 18 to 20 ohms, and as high as 37 ohms. At 37 ohms there is a reasonable match to 52 ohm coaxial cable (1.41:1), but at 25 ohms the VSWR is more than 2:1. The typical solid-state transmitter will shut down and produce little RF power at this VSWR. What is needed is a means of matching the impedance of the beam to 52 or 75 ohm coaxial cable.

The *gamma match* is shown in Figure 10.5. It consists of a piece of coaxial cable connector such that its shield is to the center point on the radiating element (L), and its center conductor goes to the matching device. The dimensions of the gamma match of Figure 10.5 are as follows:

(L is the driven element length)

$$A = \dfrac{L}{10}$$

$$B = \dfrac{L}{70}$$

Where:
  L, A and B are in meters (m).

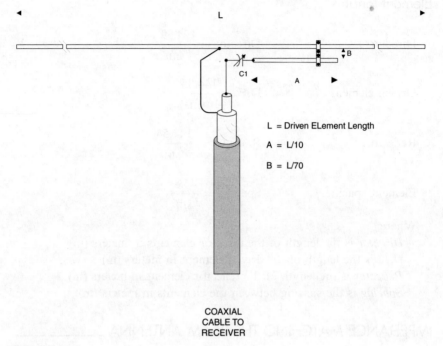

FIGURE 10.5

CHAPTER 11

Impedance matching

It is common practice to match the impedances of the antenna, transmission line, and either the receiver or the transmitter. The reason can be seen in Figure 11.1. Although the example selected is for a transmitting antenna, because of reciprocity the same notions also apply to the receive antenna, although power flows are reversed (and considerably smaller).

The case shown in Figure 11.1A is for the matched case, i.e. the transmitter output impedance, the transmission line impedance, and the antenna feedpoint impedance (which is the 'load') are all the same (e.g. 50 ohms). The thickness of the arrow is intended to represent the relative power level.

In this situation, forward power from the transmitter is sent toward the load end of the transmission line. When the radio frequency (RF) signal reaches the load it is *completely absorbed*. Some of it will go towards heating the load, and part of it will be radiated. The loss to heating in a well-designed, properly installed antenna is negligible, so it is not unreasonable to show the arrow representing the radiated portion as the same size as the forward power. The implication of the system being matched is that all of the power from the transmitter is radiated and can become a useful signal.

When the load and source are mismatched, as in Figure 11.1B, not all of the foward power is absorbed by the load. Some of it is reflected back down the transmission line toward the transmitter. When this reflected wave interferes with the forward waves still propagating up the line, the result is **standing waves** (see Chapter 2). The measure of the standing waves is the **standing wave ratio** (SWR, or VSWR if voltage is measured).

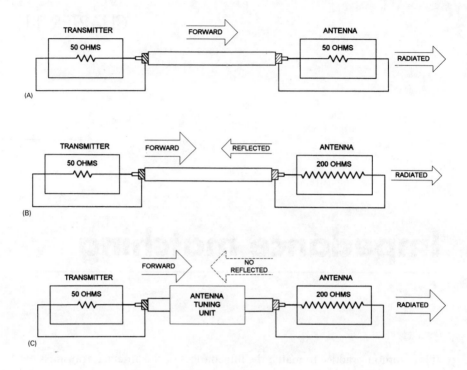

The existence of the reflected power wave has some implications in practical systems. First, it represents a loss. The reflected power is not available to contribute to the radiated signal. If no other losses occur, then this loss might prove tolerably low, but that is rarely the case. At some frequencies and in some situations, the goal is to squeeze out as much signal as possible.

Second, some transmitters, mostly modern designs with solid-state final RF power amplifiers, are intolerant of high VSWR loads. In the early days of solid-state transmitters, the first of which were CB rigs, the expensive output transistors would often blow out when even a transient high VSWR situation occurred. Older vacuum tube ('valve' to UK readers) designs were more tolerant of a high VSWR. Although those systems experienced the VSWR loss problem, the transmitter was not harmed. Modern transmitters, on the other hand, now include **automatic load control** (ALC) or other circuitry to begin reducing RF output power as the VSWR climbs. The shut down 'knee' usually starts somewhere around a VSWR of 1.5:1, and the rig will be completely shut down at some VSWR between 2.5:1 and 3:1. One rig produced 100 W of power at a VSWR of 1:1, but only 100 mW at a VSWR of 2.5:1.

A third reason is that losses in coaxial cable can be quite severe, especially at higher frequencies. Those losses increase markedly with increasing

VSWR, especially in the lower grades of coaxial cable normally used for radio antennas.

If the antenna feedpoint impedance does not match the feedline impedance, then a VSWR proportional to their ratio is created. Unfortunately, it is not always possible to create an antenna with a feedpoint impedance equal to the characteristic impedance of the feedline. Unless some means is found to effect the match, then all of the gremlins associated with a high VSWR occur.

The solution to the problem is to provide an intermediate impedance matching device (Figure 11.1C) between the transmitter (or receiver) and the antenna. The matching device might be an inductor–capacitor (LC) network, a transformer (e.g. a BALUN), or a transmission line matching section. The impedance matching device can be considered as a 'black box' between two resistances (R1 and R2, as in Figure 11.2). The property of the box is to provide the transformation that makes each resistor think it is looking at another resistor of the same value as itself. Although these devices are bilateral, i.e. the transformation occurs in both directions, it is common practice to use R1 to represent the transmitter output impedance or the receiver antenna input impedance, and R2 to represent the load impedance (e.g. antenna feedpoint impedance). In most cases, the value of R1 in transmitters is 50 ohms, and in receivers it will be either 50 ohms or 300 ohms.

LC MATCHING NETWORKS

LC matching networks are combinations of inductance and capacitance that will allow an input impedance (R1) to be transformed to an output impedance (R2). The common name for these circuits is **antenna tuner** or **antenna tuning unit** (ATU). Several common forms of ATU circuit are shown in Figures 11.3–11.5.

The circuits in Figure 11.3 are for the case where R1 > R2. This situation might occur in vertical antennas, for example, where the feedpoint impedance varies from a few ohms to about 37 ohms, both of which are less than the standard 50 ohms used by the rest of the system. The circuit in Figure 11.3A is one variant of the L-section coupler (two other variants are shown

FIGURE 11.2

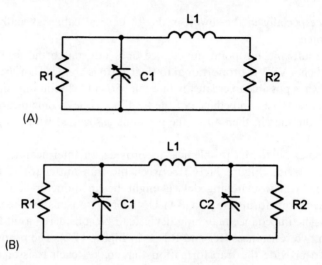

FIGURE 11.3

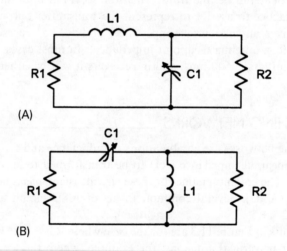

FIGURE 11.4

in Figure 11.4). In this version, the capacitor shunts the line at the input of the network, and the inductor is in series with the line.

Many homebrewed and commercial L-section couplers are designed so that changing jumpers or switch settings allows you to configure the inductor and capacitor in any of the three variants (Figures 11.3A, 11.4A, and 11.4B).

The circuit in Figure 11.4B is the π-network. This type of circuit is often used for impedance transformation when R1 is very much larger than R2 (R1 $\gg$ R2). It was once very common as a transmitter output network

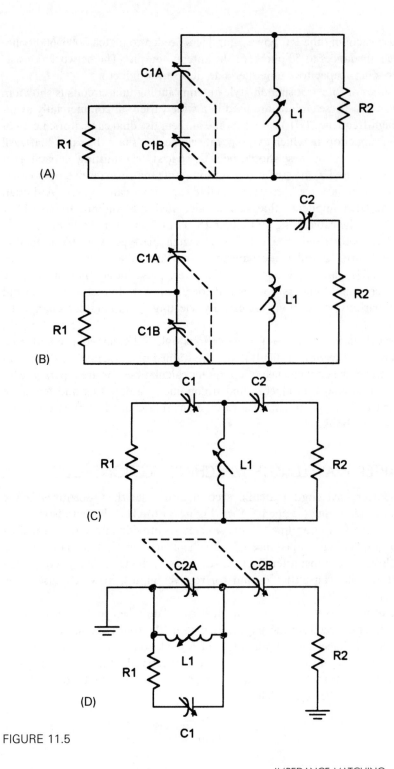

FIGURE 11.5

because vacuum tube RF power amplifiers needed to match 3000–6000 ohm anode impedances to 50 ohms for the antenna output. The network consists of two shunt capacitors on either side of a series inductor.

A series of four transmatch-style antenna tuning unit circuits is shown in Figure 11.5. These circuits are used in a wide variety of commercially available high-frequency (HF) ATUs. Some of them use dual capacitors, i.e. two-section capacitors in which two separate capacitors share the same shaft. All of the capacitors in these should be wide-spaced 'transmitting' capacitors if the ATU is used as part of a transmitting antenna. For receive-only cases, however, ordinary receiver-style variable capacitors can be used. Although the invidual component values can be calculated, it is common to use 140 or 250 pF for the capacitors, and either 18 µH or 28 µH for the inductor.

The inductor is either a variable or switchable type. Variable inductors are built with a roller mechanism that selects the turns required for any specified value of inductance. However, these inductors can be quite pricey, so many builders opt instead for an air core inductor with several taps at various points along the coil. By selecting the tap one can change the inductance.

The calculation of values for the various style of LC impedance matching network is not terribly difficult if you are trained in mathematics. But if you either prefer the convenience of computer calculations, or are a person who found a first course in algebra and trigonometry a daunting and fearsome experience, then you might wish to take advantage of the software supplement to this book.

QUARTER-WAVELENGTH MATCHING SECTIONS

The quarter-wavelength matching section, also called the Q-section, is made of transmission line (segment 'A' in Figure 11.6). The characteristic impedance of the Q-section line (Z_A) is selected to provide impedance transformation of the load impedance (Z_L) to the characteristic impedance of the line ('B') to the transmitter (Z_B). In some cases, the Q-section is connected directly to the transmitter output (or receiver input), in which case Z_L is transformed to Z_S.

There are two things that you must calculate when making a quarter-wave matching section: the **physical length of the transmission line** and the **characteristic impedance** required to match the load impedance to the source impedance. The physical length to make an electrical quarter wavelength is foreshortened from the physical quarter wavelength due to the velocity factor of the cable. The calculation is

$$L_{meters} = \frac{75V}{F_{MHz}} \text{ meters}$$

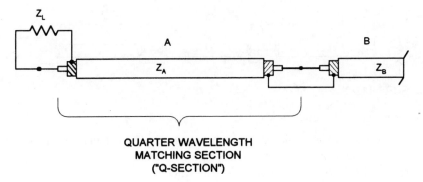

FIGURE 11.6

where L is the length in meters, F_{MHz} is the center operating frequency in megahertz, and V is the velocity factor of the transmission line (either 0.66 or 0.80 for coaxial cable – see Chapter 3).

Calculating the required characteristic impedance requires knowledge of the value of the load impedance and the characteristic impedance of the transmission line to the receiver or transmitter. The calculation is:

$$Z_A = \sqrt{Z_L Z_B}$$

For example, suppose we wish to make a Q-section for matching a 100 ohm impedance (such as found at the feedpoint of a quad loop) to 52 ohm coaxial cable. If the match is not made, then the VSWR will be about 2:1, which for many uses is clearly too high. The calculation is

$$Z_A = \sqrt{(100\,\text{ohms}) \times (52\,\text{ohms})}$$
$$Z_A = \sqrt{5200} = 72\,\text{ohms}$$

The required characteristic impedance is 72 ohms, which is a very good match to 75 ohm coaxial cable. Lest you think this is an example contrived to make the arithmetic work out to a standard impedance found in coaxial cable catalogs (which it is), it is also a very real example that is used for a number of practical antennas. Clearly, though, if the result of this equation is not near enough to some standard coaxial cable to reduce the VSWR to an acceptable level, then some other means of matching the impedance must be found.

The method of Figure 11.6 suffers from the limitations discussed above, namely the possible lack of a readily available characteristic impedance value. However, a much more general case is shown in Figure 11.7. This is the **series matching section** method. Sections L1 and L3 are made of the same type of cable, with the same characteristic impedance. The center section, L2, is made of a cable with a different characteristic impedance,

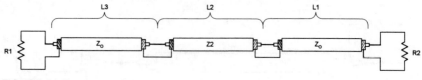

FIGURE 11.7

e.g. 75 ohm coaxial cable or 300 ohm twin-lead. The calculation of the different lengths is a bit daunting, so you are referred to the software supplement to this book in order to make the calculations.

In both the series matching section and the Q-section (which is actually a limited special case of the series section), coaxial cable, twin-lead, or parallel line can be used for the transmission lines, even though only coaxial cable is shown here.

MATCHING STUBS

Another means for effecting the impedance match is to use either a quarter-wavelength or half-wavelength **matching stub** (Figure 11.8). The stub is sometimes mounted at the feedpoint of the antenna, and the transmission line is tapped on the stub at a point (L1) that matches the impedance. In the case shown in Figure 11.8A, a shorting bar is provided at the bottom of the stub. In some cases, the stub is open and the shorting bar is not used.

Another case is shown in Figure 11.8B. In this circuit, the transmission line is connected directly to the antenna, and the stub is used to remove complex impedances by being positioned somewhere between the antenna feedpoint and a point (L1) that is a quarter wavelength down the transmission line from the feedpoint.

The final case is shown in Figure 11.8C. In this scheme, the transmission line is coaxial cable, otherwise this scheme is like Figure 11.8B. The three sections of coaxial cable are connected together using a coaxial 'tee' connector. Typically of the PL-259/SO-239 family, these connectors have three ports that are internally all connected together.

The lengths L1 and L2 depend on the VSWR experienced on the particular antenna, as well as the relationship of Z_L and Z_0. Two cases are seen:

(1) $Z_L > Z_0$:

$$L1 = \arctan \sqrt{VSWR} \text{ degrees}$$

$$L2 = \arctan\left(\frac{\sqrt{VSWR}}{VSWR - 1}\right) \text{ degrees}$$

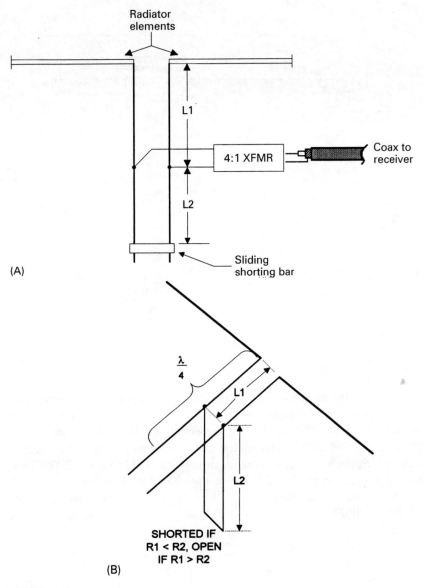

FIGURE 11.8

(2) $Z_L < Z_0$:

$$L1 = \arctan \frac{1}{\sqrt{VSWR}} \text{ degrees}$$
$$L2 = \frac{VSWR - 1}{\sqrt{VSWR}} \text{ degrees}$$

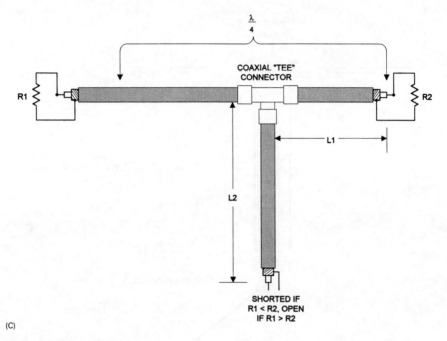

(C)

FIGURE 11.8 (*continued*)

Note that the results of these equations are in degrees. In a transmission line, one wavelength equals 360°, a half wavelength is 180°, and a quarter wavelength is 90°. You can scale the actual length against these values, keeping in mind the velocity factor of the transmission line. Consider an example. Suppose we need to match an antenna that has a 3:1 VSWR on its resonant frequency of 9.75 Mhz using 300 ohm twin-lead ($V = 0.82$). Further, the impedance of the load is 900 ohms (note: VSWR = 900/300 = 3:1). Because $Z_L > Z_0$, we use the first pair of equations:

$$L1 = \arctan \sqrt{3}$$
$$L1 = \arctan (1.73)$$
$$= 60°$$
$$L2 = \arctan\left(\frac{\sqrt{3}}{3-1}\right)$$
$$L2 = \arctan \frac{1.73}{2}$$
$$= \arctan (0.865)$$
$$= 40.9°$$

212 ANTENNA TOOLKIT

For a VSWR of 3:1, then we need L1 = 60° and L2 = 40.9°. We can calculate these as a fraction of a quarter wavelength (90°), using

$$\text{Length} = \frac{L \times 75 V}{90 F_{MHz}} \text{ meters}$$

where 'Length' is the physical length of the transmission line segment L1 or L2, L is either L1 or L2 expressed in degrees as calculated above, V is the velocity factor of the transmission line, and F_{MHz} is the operating frequency in megahertz.

For the 9.75 MHz cable, using 300 ohm twin-lead ($V = 0.82$), the physical lengths are

$$L1 = \frac{(60)(75)(0.82)}{(90)(9.75)} \text{ meters}$$

$$L1 = \frac{3690}{877.5}$$

$$= 4.2 \text{ meters}$$

$$L2 = \frac{(41)(75)(0.82)}{(90)(9.75)}$$

$$= 2.87 \text{ meters}$$

The matching stub is used on a large number of antennas. Once you get away from the simple half-wavelength dipole fed with 75 ohm coaxial cable, the number of antennas that require some form of impedance matching increases considerably, and the matching stub is used for a lot of different designs.

BALUN AND OTHER BROAD-BAND TRANSFORMERS

There are a number of different transformers used in impedance matching in antenna systems. The term 'BALUN' is used extensively, and it comes from **BALanced-UNbalanced**. Correctly used, then, the term 'BALUN' refers to a transformer that matches a balanced load (e.g. a dipole antenna feedpoint) to an unbalanced load (e.g. coaxial cable). However, it has become common (if erroneous) practice to use 'BALUN' in a generic sense to refer to any of the **broad-band transmission line transformers**. More correctly, some of those are balanced–balanced, so should be called BAL–BAL, and others are of an unbalanced–unbalanced configuration so are called UN–UN. I suppose BALUN sounds more like a word than BAL–BAL or UN–UN (note: some antenna and accessories catalogs do use these terms correctly, but the erroneous usage is also seen).

One of the earliest forms of BALUN transformer is the coaxial cable BALUN shown in Figure 11.9. Both pieces of coaxial cable used in this BALUN transformer are of the same type (e.g. 75 ohm coaxial cable). When

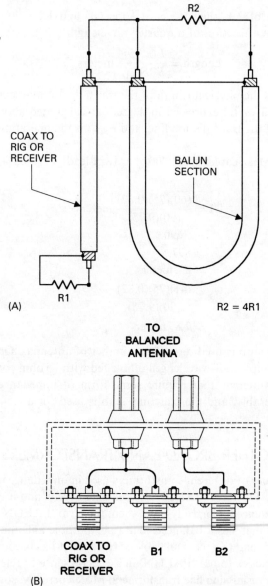

FIGURE 11.9

connected in this configuration, the BALUN transformer produces a 4:1 impedance transformation, which means that a 300 ohm balanced antenna (e.g. folded dipole) will look like a 75 ohm unbalanced load. The coaxial cable to the ham rig or receiver can be of any convenient length. The BALUN section, however, must be a half wavelength long (keeping in mind the velocity factor). The length of the BALUN section is found from:

$$L_{\text{meters}} = \frac{150V}{F_{\text{MHz}}} \text{ meters}$$

where F_{MHz} is the frequency in megahertz and V is the velocity factor (0.66 for polythene coaxial cable, and 0.80 for polyfoam coaxial cable).

A connection box for making the coaxial BALUN is shown in Figure 11.9B. This box is intended for mounting on the antenna center insulator, and should not be used for supporting the antenna unless eye bolts or other more rugged fixtures are provided at the left and right ends. The balanced antenna feedpoint is connected to a pair of five-way binding posts, while the coaxial cable for the run to the rig or receiver and the BALUN sections (B1 and B2) are connected to SO-239 coaxial connectors.

The coaxial BALUN is designed for a specific frequency, but will work over a small margin either side of the design frequency (e.g. typically one HF ham band can be accommodated). But for wide-band operation, you might want to build a broad-band transformer such as those shown in Figure 11.10. Note that some of these transformers only show one core symbol (e.g. Figure 11.10A). Those transformers have all the windings on the same core. The dots show the phase sense of the windings and indicate the same end of the winding).

The two most common forms of BALUN transformer are those in Figures 11.10A and 11.10B. The version in Figure 11.10A has no impedance transformation, and is usually referred to as a 1:1 BALUN. The transformer in Figure 11.10B, on the other hand, offers a 4:1 impedance transformation, so is equivalent to the coaxial BALUN shown in Figure 11.9A. The transformers in Figures 11.10C and 11.10D are both UN–UNs. The configuration in Figure 11.10 produces a 9:1 impedance transformation, while that in Figure 11.10D produces a 16:1 transformation.

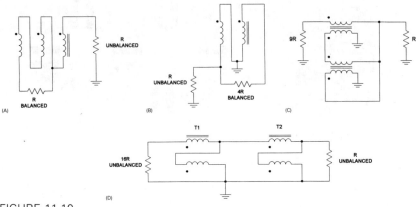

FIGURE 11.10

The construction of coil BALUNs and broad-band transformers is shown in Figure 11.11. The transformer shown in Figure 11.11A is wound on a toroid core made of either powdered iron or ferrite material. The toroid is doughnut shaped. It has the interesting attribute of containing the magnetic field to its own geometry, so has little interaction with its environment. This fact means that it will work like the book says more often than certain other transformer core configurations.

Note how the wires are wound on the toroid core. They are kept paired and lay next to each other. In this manner they are wound together as if they were only one piece. When two wires are used, this form of winding is called **bifilar winding**. When three wires are wound together in this manner, a **trifilar**

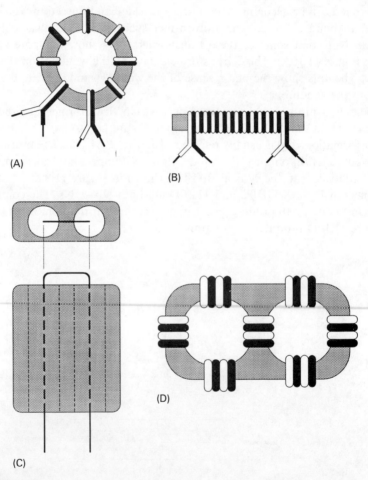

FIGURE 11.11

winding is produced. The bifilar method is used to wind the transformer in Figure 11.10B, while trifilar winding is used for that in Figure 11.10A.

The solenoid winding method is shown in Figure 11.11B. The core can be either air (in which case a coil form is needed), or a ferrite rod (as shown). Again we see the use of either bifilar or trifilar winding, depending on the nature of the transformer being made.

The so-called bazooka BALUN core is shown in Figures 11.11C and 11.11D, using two different winding styles. In the style of Figure 11.11C, the wire is passed through both holes to form a loop ('internal winding'). Counting the number of turns is a little different than one might suppose. The case shown in Figure 11.11C is one turn, even though many people erroneously assume that it is half a turn. If one end of the wire is passed through both holes one more time, then there are two turns present. Both the primary and secondary windings can be wound in this same manner, laying one over the top of the other (the primary is usually laid down first). The case shown in Figure 11.11D shows an end view of the bazooka BALUN core. In this case, several turns are wound in both the internal and external winding styles. These two styles can be intermixed on the same form, but wherever possible you are advised to use the internal winding mode preferentially.

Figure 11.12A shows how a toroid core inductor or transformer is mounted on either a printed circuit board (PCB) or metal chassis. Fiber or nylon washers are used to secure and protect the toroidal core, and nylon or other non-metallic fasteners (machine screw and hex nut) are used to keep it in place. It is important to use non-metallic fasteners to keep from interfering with the operation of the transformer. Only in the case of the largest toroids (> 5 or 6 cm in diameter) are metal fasteners usable, and even then there is some bad effect.

The scheme in Figure 11.12B is used for transmitter ATUs and similar applications where the power is higher. Two or more 5 cm or larger toroids are stacked one on top of the other. Each toroid core is first wound with fiberglass tape to insulate it from the other core. After the cores are wrapped with tape and stacked on top of each other, a final layer of tape can be added to keep the whole assembly stable. The bifilar or trifilar windings are then laid down on the stacked cores.

A number of manufacturers offer BALUNs, BAL–BALs, and UN–UNs, in both voltage and current configurations. Some are designed to replace the center insulator of an antenna such as the dipole. Others are intended for mounting elsewhere. The transformer in Figure 11.13 is intended for use on vertical antennas. The feedpoint impedance of the quarter-wavelength vertical will vary from something less than 5 ohms to about 37 ohms, depending on the installation. The transformer of Figure 11.13 is an UN–UN that has a single attachment point for 52 ohm coaxial cable to the rig or receiver.

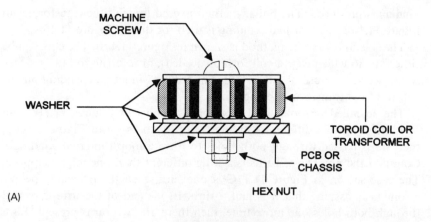

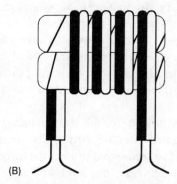

FIGURE 11.12

FIGURE 11.13

218 ANTENNA TOOLKIT

The other four coaxial connectors are various load impedances that will be transformed to 52 ohms if that port is used.

COMMERCIAL ANTENNA TUNERS

An 'antenna tuner' located at the transmitter end of the transmission line will not effect a good impedance match to the antenna, but it can mitigate the effects of a high VSWR on the transmitter. Even a single-band resonant antenna will exhibit a high enough VSWR at the ends of the band to upset the operation of the transmitter that is equipped with automatic load control (that is, all modern solid-state rigs). Although antenna tuner construction is well within the abilities of most do-it-yourself 'homebrewers,' it is also likely that most will opt instead for a ready-built ATU. A number of reasonably priced models are available on the market.

Figure 11.14 shows the MFJ-956 receiver S/M/L wave preselector/tuner. This device places a series resonant circuit in series with the antenna downlead before it goes to the receiver. A single variable capacitor (365 pF maximum) is in series with an inductance. Four different switch-selected inductance values are provided (2.5 mH, 330 µH, 22 µH, and 1 µH) to cover the range 150 kHz to 30 MHz. When the inductor switch is in the BYP ('bypass') position, the preselector/tuner is effectively out of the circuit, and the antenna–receiver combination works as normal. In the GND ('ground') position, the antenna input of the receiver (but not the antenna) is grounded.

Tuners for transmitting use are shown in Figures 11.15 and 11.16. The tuner in Figure 11.15 is the MFJ-949E Deluxe Versa Tuner II, while that in Figure 11.16 is the MFJ-986 Differential Tee tuner. These tuners are a bit larger than the standard receiver-only tuner, but are nonetheless just as

FIGURE 11.14

FIGURE 11.15

FIGURE 11.16

useful to the receiver owner as well. The only thing that will not work for receiver owners is the VSWR metering system. These are based on the same technology as RF wattmeters, so require the output of a radio transmitter for excitation. Both models allow more than one antenna (although only one at a time!) to be switch selected.

CHAPTER 12

Simple antenna instrumentation and measurements

It is very difficult to get an antenna working properly without making some simple measurements. Although it would be nice to make pattern measurements, those are beyond the reach of almost all of us. On the other hand, the 'I worked [or heard] a guy on the other side of the world' type of measurement tells us little or nothing. During the peak of the sunspot cycle a breath of hot air on the antenna can be picked up on all continents.

There are some things that can be measured, however. For example, the voltage standing wave ratio (VSWR) and the resonant frequency of the antenna are readily accessible. It is also possible to measure the impedance of the antenna feedpoint. Hams can measure the VSWR either with a special VSWR meter (often built into transmitters or antenna tuning units), or by using a radio frequency (RF) wattmeter.

By stepping through the band and testing the VSWR at various frequencies, one can draw a VSWR curve (Figure 12.1) that shows how the antenna performs across the band. The resonant frequency is the point where the VSWR dips to a minimum. You can use the resonant frequency to figure out whether the antenna is too long (resonant frequency lower than the hoped-for design frequency), or too short (resonant frequency above the design frequency).

But resonant frequency and VSWR curves are not the entire story because they do not tell us anything about the impedance presented by the antenna. One cannot get the VSWR to be 1:1 unless the antenna impedance and transmission line impedance are the same. For example, a dipole has a nominal textbook impedance of 73 ohms, so makes a very good match

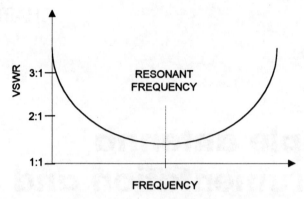

FIGURE 12.1

to 75 ohm coaxial cable. But the actual impedance of a real dipole may vary from a few ohms to more than 100 ohms. For example, if your antenna exhibits a feedpoint impedance of (say) 25 ohms, using 75 ohm coaxial cable to feed it produces a VSWR = 75/25 = 3:1. Not too great. Measuring the feedpoint impedance is therefore quite important to making the antenna work properly.

There are any number of instruments on the market that will aid in making antenna measurements. Some of them are quite reasonably priced (or can be built), while others are beyond the reach of all but the most ardent and well-endowed enthusiasts. In this chapter we will look at RF power meters, VSWR meters, dip meters, noise bridges, impedance bridges, and a newer breed of more universal instrument called the SWR analyzer. We will also look at a few accessories that are useful in the measurement or adjustment of radio antennas (e.g. coaxial switches, attenuators, and dummy loads).

RF POWER METERS AND VSWR METERS

The RF power meter is designed to indicate the actual value of RF power (in watts). VSWR meters are similar, but are either calibrated in VSWR units or have a selectable scale that is so calibrated. Some RF wattmeters measure the voltage on the transmission line, while others measure the current flowing in the line. If things are done right, then these approaches work nicely. But sometimes those methods can be fooled. A better system is to measure both the current and the voltage and to display their product. Although this task sounds like a formidable instrumentation job, it is actually the way one of the most popular RF power meters works. The Bird Electronics Model 43 (Figure 12.2) is one of the oldest and most popular professional-grade instruments. It uses a specially designed internal transmission line segment

FIGURE 12.2

with sampling loops inserted in the form of plug-in 'elements' (the round object with an arrow in Figure 12.2). The elements are customized for the RF power level and frequency range. The arrow on the element indicates the direction of the power flow. When the arrow points to the antenna feedline, the instrument is measuring the forward RF power, and if pointed toward the transmitter it is measuring the reflected RF power.

Although the Model 43 type of meter will provide the RF power measurements more accurately under a wider range of conditions, a ham radio style VSWR meter/RF power meter will work well (and measure accurately) under most ham conditions. In addition, these instruments will provide a measure of the VSWR (with an RF wattmeter you need to calculate the VSWR). Figure 12.3 shows two typical ham-style VSWR/RF power meters. The instrument in Figure 12.3A is small and simple, but quite effective. It will measure forward and reflected power over the entire high-frequency (HF) band, as well as the VSWR. To measure the VSWR, set the full-scale/SWR switch (marked 'FS' and 'SWR') to the FS position, key the transmitter, and adjust the knob to achieve a full-scale deflection of the meter pointer. When the FS/SWR switch is set to the SWR position, it will read the VSWR. These instructions, by the way, fit a large number of different instruments, by many manufacturers, in this same class.

The instrument in Figure 12.3B is for the very high/ultrahigh frequency (VHF/UHF) bands. It does not have to be set as does the meter above. The

FIGURE 12.3

reason is that it uses the 'crossed-needles' method of displaying forward and reverse power. Lines of intersection with the two scales form a species of nomograph that reads out the VSWR from observation of the point where the needles cross. This instrument also includes a **peak reading** function. Most passive RF power meters inherently measure the average RF power because of the inertia of the meter movement. But with active circuitry, i.e. circuits that require direct-current power to operate, it is possible to sample and hold the peaks of the RF waveform. This meter then becomes more useful for measuring continuous-wave and amplitude-modulated waveforms (including SSB). Some authorities claim that the peak reading meter is preferred for SSB operation – and I concur.

DIP METERS

The **dip meter** works because a resonant circuit will draw power from any nearby field oscillating at its resonant frequency. An antenna that is cut to a specific frequency is resonant, so acts like an LC resonant circuit. At one time, these instruments were called **grid dip meters** because they were vacuum tube ('valve') operated. The metering monitored the grid current to detect the sudden decrease when the instrument was coupled to an external resonant circuit. Today, some people still use the archaic term 'grid dipper' even though dip meters have not had grids for nearly 30 years.

Figure 12.4A shows a typical dip meter. It consists of an LC oscillator in which the inductor is protruding outside the box so that it can be coupled to the circuit under test. A large dial is calibrated for frequency, and a meter monitors the signal level. Coils for additional bands are stored in the carrying case.

The frequency dial of the dip meter bears some additional comment. It typically covers a very wide range of frequencies, yet the dip we need is very sharp. The calibration is none too accurate. As a result, it is wise to find the dip, and then use a receiver to measure the operating frequency while the dip meter coil is still coupled to the circuit being tested. This operation is a neat trick unless you are possessed with good manual dexterity. It is necessary to measure the frequency while the instrument is coupled because this type of LC oscillator changes frequency markedly when the loading of the coil changes.

The proper method for coupling the dip meter to an antenna transmission line is shown in Figure 12.4B. Construct a 'gimmick', i.e. a small coil of one to three turns, of sufficient diameter to slip over the coil protruding from the dip meter. Either slip the dipper coil inside the gimmick, or bring it into close proximity. Energy is transferred from the dipper to the gimmick and thus to the antenna. When the dipper is tuned to the resonant frequency of the antenna, a sudden drain is put on the dipper, and the signal level drops sharply. The frequency of this dip tells you the resonant frequency of the antenna.

The dipper must be tuned very slowly – very, very slowly – or you will miss the dipping action of the meter. Tuning too fast is the principal reason why many newcomers fail to make the dipper work. To make matters more confusing, the meter needle will drift all over the scale as the instrument is tuned. This occurs because it is natural for LC-tuned oscillators to produce different signal level at different frequencies (generally, but not always, decreasing as frequency increases). The effect to watch for is a very sharp drop in the meter reading.

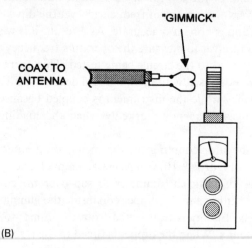

FIGURE 12.4

ANTENNA IMPEDANCE (OR 'RESISTANCE') BRIDGES

While it is necessary to measure the VSWR of an antenna to see how well it is working, the VSWR alone is insufficient to optimize performance. If you do not know the feedpoint impedance at resonance (minimum VSWR), then you cannot do much to correct a problem. Several instruments are available that will measure the feedpoint resistance of the antenna (it is called 'impedance' in some user manuals, but it is really only the resistive component of impedance). The instruments in Figure 12.5 are examples of antenna resistance meters. The instrument in Figure 12.5A is based on the Wheatstone bridge circuit. It uses a variable resistor (connected to the thumbwheel resistance dial) as one leg of the bridge, and the antenna

(A)

(B)

FIGURE 12.5

resistance another leg (fixed resistors form the remaining two legs of the classic Wheatstone bridge). This model is passive, and requires an external signal source. There is an internal amplifier that allows you to use a signal generator, or the amplifier can be bypassed if you wish to drive it with a transmitter signal.

The instrument in Figure 12.5B is a little different. It is basically a sensitive VSWR meter with a built-in signal generator. Knowing the system impedance it can thereby calculate the resistance from knowledge of the VSWR.

RF NOISE BRIDGES

The **noise bridge** (Figure 12.6A) is an interesting antenna instrument that is used in conjunction with a radio receiver to measure the antenna feedpoint resistance, approximate the reactance component of the impedance, and find the resonant frequency. These instruments use white noise ('hiss' is what it sounds like on the receiver speaker) to generate a wide spectrum of radio frequencies. When the receiver is used to monitor the noise, a null is noted at the resonant frequency.

Figure 12.6B shows how the noise bridge is connected between the antenna transmission line and the antenna input connector of the receiver. The line between the bridge and the receiver can be any length, but it is wise

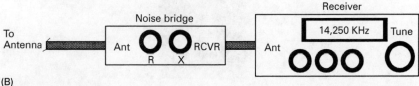

FIGURE 12.6

to keep it short. One reader of my columns wrote in to take exception to the 'short as possible' requirement because it is not strictly necessary in a *perfect world*. My own experience says differently, especially at higher frequencies where the line is a significant fraction of a wavelength. A meter or so should not affect any HF measurement, however.

The line to the antenna should either be as short as possible (preferably zero length), but in the real world, where one has to be more practical than theoretically pure, it is sufficient to make the transmission line an integer multiple of an electrical half wavelength long (the physical length is shorter than the free space half wavelength because of the velocity factor). The impedance hanging on the far end of the half wavelength $\times N$ line is repeated every half wavelength. A consequence of this effect is that the impedance measured at the transmitter or receiver end is the same as at the antenna end.

To use the noise bridge, set the resistance (R) and reactance (X) controls to mid-scale (X = 0). Adjust the receiver frequency to the expected resonant frequency. At this point you can do either of two different measurement schemes.

Method 1

First, you can set the R-control of the noise bridge to the desired antenna impedance, and then tune the receiver until a dip in the noise occurs (as heard in the output or noted on the S-meter). For example, suppose you have a dipole cut for 11.75 MHz, and installed such that the expected feedpoint impedance should be around 70 ohms. It is fed with a half wavelength of polyfoam dielectric 75 ohm coaxial cable ($V = 0.80$). The cable length is $(150)(0.8)/11.75 = 10.2$ m. The noise bridge X-control is set to 0, and the resistance control is set to the letter that represents 70 ohms (given in the calibration manual). This resistance setting can also be determined experimentally by connecting a 70 ohm resistor across the antenna terminal and adjusting the R-control for the minimum noise. When the noise bridge is set, then adjust the receiver for minimum noise. The frequency at which the dip occurs is the actual resonant frequency. I used this method to investigate a vertical cut for 21.25 MHz, only to find the actual resonance was 19.2 MHz, which explained the high VSWR.

Method 2

The other method is to set the receiver to the design resonant frequency of the antenna. You then adjust the R-control for a dip in the noise level. The value of the resistance is the feedpoint impedance of the antenna. You can either match that impedance, or adjust the antenna to bring the actual resonant frequency closer to the design frequency.

VSWR ANALYZERS

The basic premise in this section is that the instruments used must be accessible to people who do not have a ham radio or professional radio operator's license. Some of the instruments discussed above meet that requirement, but a relatively new breed of instrument called the SWR analyzer provides a lot of capability to the short-wave listener, scanner operator, and ham radio operator alike. It uses a low-power RF signal generator and some clever circuitry to measure the VSWR of the antenna. One model also measures the feedpoint resistance.

The simple version shown in Figure 12.7 is for the low VHF band (up to and including the 6 m ham band). It is a hand-held, battery-powered instrument. The meter reads the VSWR of the antenna at the frequency set by the TUNE dial.

A somewhat more sophisticated instrument, the MFJ Enterprises Model MFJ-259, is shown in Figure 12.8. The front panel has two meters, SWR and RESISTANCE. The SWR meter is calibrated up to 3:1, with a little uncalibrated scale to indicate higher SWRs. The RESISTANCE meter is calibrated from 0 to 500 ohms, which is consistent with the SWR range. Two controls on the front panel are TUNE and FREQUENCY (MHz) (a

FIGURE 12.7

FIGURE 12.8

bandswitch). The MFJ-259 has a digital frequency meter to measure the operating frequency of the internal oscillator. This frequency counter can also be used to measure the frequency of external signal sources (DO NOT connect the counter to the output of a transmitter – the instrument will be destroyed). The top end of the MFJ-259 has a number of controls and connectors. An SO-239 'UHF' style coaxial connector is provided for the antenna connection. A BNC coaxial connector is provided to apply an external signal to the frequency counter, while a pushbutton INPUT switch is available to switch the counter from internal to external signal sources. Another pushbutton switch is used to set the gate timing of the counter (a red LED on the front panel blinks every time the gate is triggered). The tuning is from 1.8 to 174 MHz, while the counter will measure up to 200 MHz.

The MFJ-259 will work from an external 12 V DC source, or from an internal battery pack consisting of eight size-AA standard cells. MFJ recommends that either alkaline or rechargeable batteries, rather than ordinary zinc–carbon cells, be used in order to reduce the possibility of leakage that could damage the instrument (this is good practice in all battery-powered instruments). I have a homebrewed battery pack that uses eight size-D nickel cadmium batteries (4 A-h rating) that can be recharged from a 12 V DC power supply, and it works well with the MFJ-259.

Unlike many lesser SWR meters, this instrument is not fooled by antennas that have impedances consisting of both resistance and reactance elements. An example in the manual demonstrates an impedance of 25 + j25 ohms (i.e. R is 25 ohms and reactance, X, is also 25 ohms). When connected to a 50 ohm load one might be tempted to think the VSWR is 1:1, and some cheaper meters will so indicate. But the actual SWR is 2.6:1, which is what the MFJ-259 will read.

The resistance measurement assumes a resistive load (i.e. the measurement is made at the resonant frequency of the antenna), and is referenced to 50 ohms. The VSWR and resistance measurements should be consistent with each other. If the VSWR is 2:1, then the resistance should be either 100 ohms (100/50 = 2:1) or 25 ohms (50/25 = 2:1). If the resistance is not consistent with the VSWR reading, then you should assume that the impedance has a significant reactive component, and take steps to tune it out.

In addition to antenna measurements, the MFJ-259 is equipped to measure a wide variety of other things as well. It will measure the velocity factor of transmission line, help in tuning or adjusting matching stubs or matching networks, measure capacitance or inductance, and measure the resonant frequency of LC networks.

Figure 12.9 shows an MFJ-249 meter (similar to the MFJ-259, but without the resistance measurement) equipped with the MFJ-66 dip meter adapter. It can be used to make the MFJ-249 or MFJ-259 work in the same manner as a dip meter. Using this adapter allows you to measure the resonant frequency of tank circuits using the dipper approach, as well as to measure things such as the coefficient of coupling between two LC circuits, transformers, and other radio circuits.

DUMMY LOADS (ARTIFICIAL AERIAL)

The **dummy load**, or **artificial aerial** as it is also called, is a resistor used to simulate an antenna when adjusting a transmitter or other radio apparatus. These devices consist of a non-inductive, 50 ohm resistor inside a shielded enclosure.

The dummy load offers several benefits over regular antennas. First, it is a constant resistance over the entire frequency range, and, second, it does not present any appreciable reactance. Third, and perhaps most important, the dummy load provides the ability to conduct tests off the air where you will not cause television interference, broadcast interference, or interference to other users of the test frequency that you select. Besides, it is illegal and just plain rude to radiate when you do not have to!

One use for dummy loads is to adjust antenna tuners. You can connect the dummy load to the antenna output of the tuner, and then adjust the

FIGURE 12.9

controls for best VSWR at a number of frequencies. By recording the knob settings, you can 'rough in' the antenna tuner when changing frequency.

Several types of dummy load are shown in Figures 12.10 and 12.11. The version in Figure 12.10 is a small 5 W HF model intended for measuring the output of CB transmitters. I have used this particular dummy load for adjusting a wide range of RF projects and instruments over the years. To make accurate power measurements, the dummy load is placed at the ANT connector of the RF power meter, and the transmitter connected to the other end. The power reading is not obscured by transmission line effects or the reactance that a real antenna might present.

The dummy loads shown in Figure 12.11 are intended for ham radio power levels. The model in Figure 12.11A is air cooled, and operates at RF power levels up to 1500 W, over a range of 1–650 MHz. The version in Figure 12.11B is oil cooled, and will handle up to 1000 W of RF power.

COAXIAL SWITCHES

A coaxial switch (Figure 12.12) is used to allow a receiver or ham radio set to use any of several antennas (models with up to 16 ports are available, but this one is a two-port model). The common connector is for the receiver or transmitter, while the two antennas are connector to the 'A' and 'B' ports,

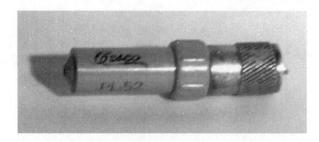

FIGURE 12.10

FIGURE 12.11

234 ANTENNA TOOLKIT

FIGURE 12.12

respectively. Alternatively, one can turn the switch around backwards (it is bidirectional, after all), and use the same antenna on two different receivers or transmitters.

The use of the coaxial switch in antenna measurement is in comparing the antenna being tested with either another antenna or a dummy load. The kind of off-the-air checks that amateurs and short-wave listeners can make are notoriously inaccurate, but can be made a lot more useful by making comparisons with known antennas. For example, a friend of mine, the late Johnnie H. Thorne (K4NFU/5) had an antenna farm in Texas (and it did seem that he grew antennas, judging from the number he had). He kept a standard dipole, optimally installed and cut for 20 m, and made all of his test designs for the same frequency. He would compare new designs to the dipole by switching back and forth while monitoring the signal strength on the receiver S-meter. He could also compare two different antennas by comparing them against each other or against the dipole.

STEP ATTENUATORS

The **step attenuator** (Figure 12.13) is a precision instrument that provides highly accurate levels of attenuation in steps of 1 dB, 2 dB, 3 dB, 5 dB, 10 dBV, and one or more 20 dB settings. You can use these to calibrate instruments, or measure signal levels. One use is to make comparison measurements of two signal sources, two antennas, etc.

For example, suppose you want to measure the gain of a new antenna relative to a dipole (assuming that both are cut for the same frequency). You would pick or provide a distant signal on the resonant frequency of the antenna. The attenuator is inserted into the feedline path of the test antenna (Figure 12.14). The signal strength is then measured with the dipole in the

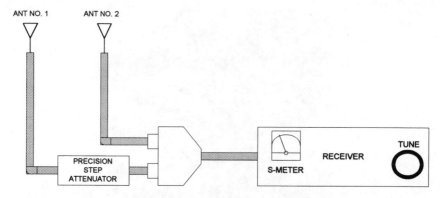

FIGURE 12.13

FIGURE 12.14

circuit (see the coaxial switch discussion above), and the reading of the S-meter noted. If you have an RF sensitivity control, then adjust it to a convenient point on the meter dial, like S-9 or some other marked spot. Next, cut to the new antenna and note the signal strength. If it increased, then select the various levels of attenuation until the new signal level is the same as the initial signal level. The set level of attenuation is equivalent to the gain difference between the antennas. If the new signal level is lower than the initial signal level, then the new antenna has less gain than the old antenna. Put the attenuator in line with the comparison antenna, rather than the test antenna.

CHAPTER 13

Getting a 'good ground'

Because this book is about radio antennas it is necessary to also spend some time on the topic of antenna grounds. The effectiveness and safety of a radio antenna often hangs on whether or not it has a good radio frequency (RF) ground. Poor grounds cause most forms of antenna to operate at less than best efficiency (sometimes a whole lot less). In fact, when transmitting it is possible to burn up between 50 and 90% of your RF power heating the ground losses under the antenna, instead of propagating into the air. Ground resistances can vary from very low values of 5 ohms, up to more than 100 ohms (5–30 ohms is a frequently quoted range). RF power is dissipated in the ground resistance, so is not available for radiation. The factors that affect the ground resistance include the **conductivity of the ground**, its chemical **composition**, and its **water content**.

The ideal ground depth is rarely right on the surface, and, depending on the local water table level, might be a couple of meters or so below the surface. It is common practice among some amateur radio operators to use the building electrical ground wiring for the RF antenna ground of their station. Neglecting to install an outdoor ground that will properly do the job, they opt instead for a single connection to the grounded 'third wire' in a nearby electrical outlet. Besides being dangerous to work with, unless you know what you are doing, it is also a very poor RF ground. It is too long for even the low RF bands, and radiates RF around the house in large quantity. Stations that use the household electrical wiring as the radio ground tend to cause television interference, and other electromagnetic interference both in their own building and in nearby buildings.

Other people use the plumbing piping in the house, but this is also a problem. While the cold water pipe entering older houses is metal, and is buried for quite a distance underground, it is also a single conductor, and is often made of a metal that is not optimal for radio work. Also, in modern buildings, the main water service and the distribution pipes in the plumbing system of the building may well not be metal at all, but rather PVC or some other insulating synthetic material. Even if you get a metal water pipe, if it is the hot water pipe then no effective ground exists. There is also a very severe danger of hooking your ground to the gas service line, which could *cause a fire or explosion*. In the USA, gas pipes are usually painted black, or are unpainted, but enough water pipes also fit this description that it is good practice to avoid the pipes in a building when searching for a radio grounding solution.

TRADITIONAL RADIO GROUNDING SOLUTIONS

The goal in making a radio ground is to provide a low resistance path to the earth for radio signals. An additional goal is to make a path to the earth for lightning that strikes the antenna, or even nearby (see Chapter 2 for the use of lightning arrestors in antenna lines).

We can reduce the ground resistance by either altering the composition of the earth surrounding the ground point, or by using a large surface area conductor as the ground point. Figure 13.1 shows the traditional ground rod used by small radio stations, including amateur and receive-only stations. Use either a copper (or copper-clad steel) rod at least 2–3 m long. Electrical supply houses, as well as amateur radio and communications equipment suppliers, also sell these ground rods. Do not use the non-clad steel types sold by some electrical supply houses. They are usable by electricians when making a service entrance ground in your home or workplace, but RF applications require the low skin resistance of the copper-clad variety. The rod need not be all copper, because of the skin effect forcing the RF current to flow only on the outer surface of the rod. Try to use a 2.5 m or longer rod if at all possible, because it will work better than the shorter kind. Do not bother with the small television antenna 100–150 cm ground rods; they are next to useless for high-frequency (HF) radio stations. Drive the ground rod into the earth until only a few centimeters remain above the surface. Connect a heavy gage ground wire from your station to the ground rod (some people use heavy braid or the sort once used for the return connection from the battery to the body of an automobile; an additional source is salvaged braid from the heavier gages of coaxial cable). The ground wire should be as short as possible. Furthermore, it should be a low-inductance conductor. Use either heavy braid (or the outer conductor stripped from RG-8 or RG-11 coaxial cable), or sheet copper. You can buy

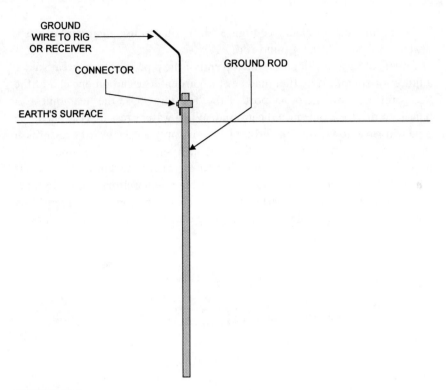

FIGURE 13.1

rolls of sheet copper from metal distributors in widths from 10 cm up to about 50 cm. Some amateurs prefer to use 17.5 cm-wide roofing foil that is rated at a weight of 1.5 kg/m. Sweat solder the ground wire to the rod. You can get away with using mechanical connections like the electricians use, but will eventually have to service the installation when corrosion takes its toll. I prefer to use soldered connections, and then cover the joint with either petroleum jelly or acrylic spray lacquer.

Another alternative is to use a copper plumbing pipe as the ground rod. The pipe can be purchased in 2–5 m lengths from plumbing supply shops or hardware stores. The pipe selected should be 19 mm in diameter or larger. Some people report using up to 50 mm diameter pipe for this application. The surface area of the hollow pipe is greater than that of a solid rod of the same diameter. Because of certain current flow geometries in the system, however, the ground resistance is not half the resistance of a rod the same diameter, but is nonetheless lower.

Unlike the copper-clad steel rod, the copper pipe has no compression strength and will deform when hit with a hammer or other driving tool.

To overcome this problem (see Figure 13.2) you can use a garden hose as a 'water drill' to sink the ground rod.

Sweat solder a 'tee'-joint on the top end of the pipe, and then sweat solder a faucet or spigot fitting that matches the garden hose end on one end of the 'tee'-joint. Cap off the other port of the 'tee'-joint. Use the 'tee'-joint as a handle to drive the pipe into the ground. When water pressure is applied, the pipe will sink slowly into the ground as you apply a downward pressure on the tee-handle. In some cases, the pipe will slip into the ground easily, requiring only a few minutes. In other cases, where the soil is hard or has a heavy clay content, it will take considerably greater effort and more time. When you finish the task, turn off the water and remove the garden hose. Some people also recommend unsoldering and removing the 'tee'-joint.

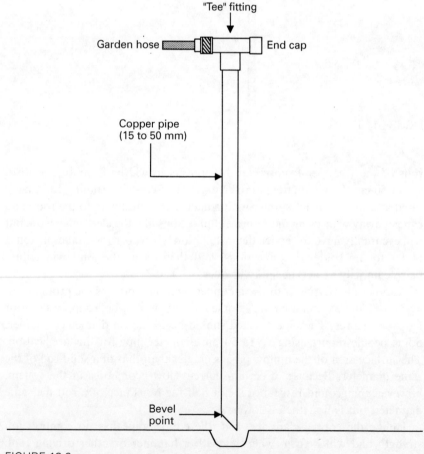

FIGURE 13.2

ALTERING SOIL CONDUCTIVITY

The conductivity of the soil determines how well, or how poorly, it conducts electrical current (Table 13.1). Moist soil over a brackish water dome conducts best. In the USA, the southern swamps make better radio station locations than most of the rest of the country; and the sand of the western deserts make the worst.

In Arizona, for example, one finds that few buildings have cellers or basements. The reason is that there is a thick calcium-based layer (called 'caleche') a meter or two down that resists even mechanized diggers. To build homes in that sand they have to drill holes in the caleche and then sink pylons down to bedrock.

The same caleche also prevents water from sinking into the sand, so it flows off the land in natural channels in the desert. The area is rather dangerous for out-of-state campers who do not understand that those nice flat, areas at the bottom of gulleys have a habit of filling in seconds if there is a distant thunder storm in the mountains. They look up, and to their horror see a 3–5 m high wall of water rushing toward them at 50 km/h! During the late summer months, Arizona has a 'monsoon' season when it rains hard and floods every day (90% of the state's annual rainfall comes in a 6-week period from sometime in July to mid-September). Evaporation is so rapid that an hour after the rain stops, there is little evidence of the rainfall except in the gulleys. If you have never felt hot rain, then visit southern Arizona in August – but be prepared for temperatures around 43°C in the shade and 80°C on the front seat of an automobile if you are so unwise as to leave all the windows rolled up.

Figure 13.3 shows a method for reducing the soil electrical resistance by treating with one of three chemicals: copper sulfate, magnesium sulfate or common rock salt. Rock salt is one of several salt materials used for snow and ice melting in snow-prone areas. If you cannot locate rock salt in a hardware store, then look for a store that sells ice cream making supplies. Rock salt is a principal ingredient in the process (but not the product).

TABLE 13.1

Type of soil	Dielectric constant	Conductivity (siemans/m)	Relative quality
Salt water	81	5	Best
Fresh water	80	0.001	Very poor
Pastoral hills	14–20	0.03–0.01	Very good
Marshy, wooded	12	0.0075	Average/poor
Rocky hills	12–14	10	Poor
Sandy	10	0.002	Poor
Cities	3–5	0.001	Very poor

Figure 13.3A shows a side sectioned view of a circular slit trench method of applying the chemical treatment to reduce ground resistance. **WARNING**: before using this method, you need to determine whether or not you really want to do it. The method involves putting conductive chemicals into the ground. *Make sure that local, state or national environmental regulations do not prohibit the practice.* Also, if you ever want to use the spot to grow vegetables or flowers – or even grass – it will prove difficult because the chemicals essentially poison the soil for several years. Fortunately, they are water-soluble, so the effect will not last forever.

Dig a 15–30 cm deep circular trench, about 30–60 cm from the ground rod. Fill that trench with a 10–12 cm layer of rock salt, magnesium sulfate, or copper sulfate (use only one). Water the trench well for about 15 minutes. Cover the remaining depth with soil removed when you have dug the trench. If you use about 100 kg of material, the treatment will need to be repeated every 24–36 months, depending upon the local rainfall and soil composition.

Figure 13.3B shows an alternative method: the saltpipe. Use either copper or PVC plumbing pipe, up to 10 cm in diameter (although 4–8 cm is easier to work). The overall length of the pipe should be at least 50–150 cm. Although longer pipes are useful, they are also difficult to install. Drill a large number of small holes in the pipe (no hole over 10 mm), sparing only the end that will be above the surface. Cap off both ends.

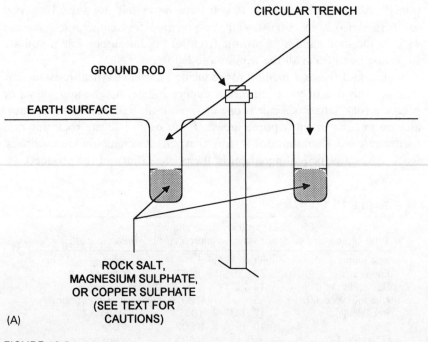

FIGURE 13.3

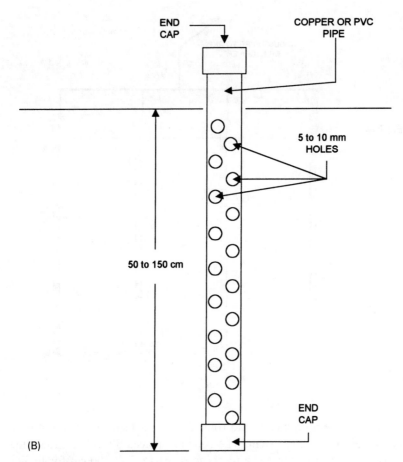

(B)

FIGURE 13.3 (continued)

Install several saltpipes in a pattern around the ground rod. Installation is best accomplished using a fencepost hole digger. Drop the saltpipe into the hole and backfill with water. Remove the top endcap and hose down for about 15 minutes, or until all the material is gone. Refill the pipe occasionally to account for the salt leaching out. When the chemical is completely leeched out of the pipe, refill the pipe and recap the top end. Leave the pipe in place.

Be certain of local environmental laws before using the above methods.

MULTIPLE GROUND RODS

The key to a low-resistance ground is the surface area in contact with the soil. One means for gaining surface area, and thereby reducing ground

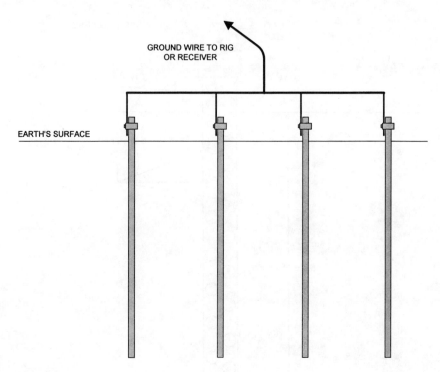

FIGURE 13.4

resistance, is to use multiple ground rods. Figure 13.4 shows the use of four ground rods in the same system. The rods are placed 30–100 cm apart for low- and medium-power levels, and perhaps more than 100 cm for higher amateur radio power levels. An electrical connection is made between the rods on the surface using either copper stripping or copper braid. The connections are sweat soldered in the manner described above, with the feedpoint at the center rod.

RADIALS AND COUNTERPOISE GROUNDS

The effectiveness of the ground system is enhanced substantially by the use of radials either above ground, on the surface, or buried under the surface. Figure 13.5 shows a vertical antenna with a set of ground radials. It is not unreasonable to use both radials and a ground rod. Vertical antennas are relatively ineffective unless provided with a good ground system, and for most installations that requirement is best met through a system of ground radials.

An effective system of radials requires a large number of radials. Although as few as two quarter-wavelength resonant radials will provide

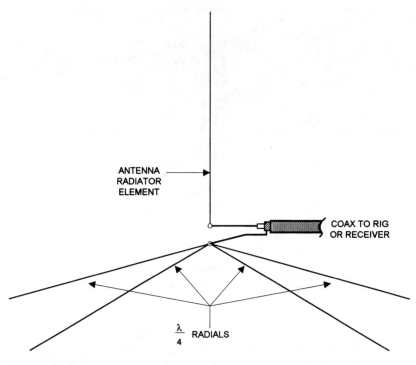

FIGURE 13.5

a significant improvement, the best performance is to use more. Broadcasters in the AM band (550–1640 kHz) are advised to use 120 half-wavelength radials. Installing more than 120 radials is both expensive and time-consuming, but does not provide any substantial improvement. For amateur and small commercial stations, a minimum of 16 quarter-wavelength radials should be used. Above ground, the use of insulated wire is recommended to reduce the possibility of anyone receiving RF burn if it is accidentally touched while transmitting. Although some sources claim that any size wire from AWG 26 up to AWG 10 (SWG 0–12) can be used, it is best to use larger sizes in that range (i.e. AWG 14 through AWG 10 (SWG 16–12)). Either solid or stranded wire can be used.

When viewed from above, the radials should be laid out in a symmetrical pattern around the antenna. This coverage provides both the lowest resistance and the best radiation pattern for the antenna. Solder all radials together at a common point, which might be the ground or mounting rod used to support the vertical antenna.

If your antenna is ground mounted, or if the radials are used with the station ground, then use a system such as that in Figure 13.6 to make the connections. A clamp on the ground rod is used to bind together the rod, the radials and the ground wire from the receiver or transmitter. The radials

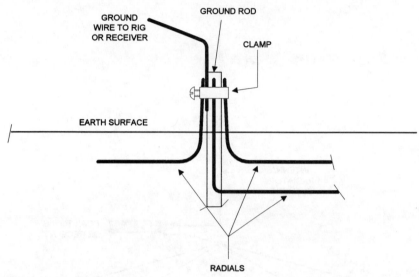

FIGURE 13.6

should be buried a few centimeters below the surface. Some people report using the tip of a spade or shovel to dig a very thin trench into which the wire can be pushed. Be careful to get the radial wire deep enough. Ordinary soil dynamics can make the wire rise to the surface within a year or so.

TUNING THE GROUND WIRE

An alternative that some operators use is the **ground wire tuner**. These instruments insert a series-tuned inductor–capacitor (LC network) in series with the ground line. These devices are usually called **artificial ground tuners**, and are inserted into the ground line of the station (Figure 13.7). Ground line tuners are especially recommended for cases where the wire to the ground is too long, in the case of single-wire antennas (e.g. Windom antennas or random-length wires), or anytime the tuning seems peculiar, but the antenna tuning unit works well into a dummy load. The purpose of the unit is to tune out any inductive or capacitive reactance on the line. Transmitter operators adjust the ground line tuner for maximum ground current at the operating frequency, although receiver operators will have to experiment a bit to find the correct settings. MFJ Enterprizes, Inc. (PO Box 494, Mississippi State, MS 39762, USA) makes several such devices. The version shown in Figure 13.8 is a stand-alone unit that contains only the ground current meter, inductor, and capacitor. The instrument in Figure 13.9 is a combination unit that includes both the ground line tuner and antenna tuner in one box.

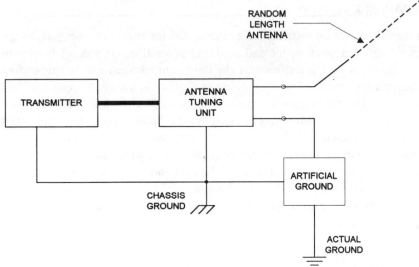

FIGURE 13.7

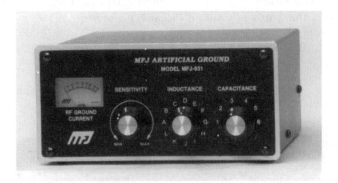

FIGURE 13.8

FIGURE 13.9

GETTING A 'GOOD GROUND' **247**

WARNING

Grounds for radio installations are intended for two purposes: making the radio antenna work better and lightning protection. Be absolutely sure to check with local authorities over the minimum requirements for grounding antenna structures – and then doing more than is minimally required. Not only is it possible to get into trouble with the authorities with a poor ground, you may also find that your homeowners and liability insurance is not valid if an uninspected installation is made. My own homeowners policy states that the insurance coverage is not in effect if an antenna is erected that meets the minimum requirements for a mechanical and electrical inspection by the county government but does not have a valid inspection certificate. Check with your local government and insurer before building any antenna or ground system.

CONCLUSION

Now that you have learned how to build safe and efficient antennas and grounds, in the words of Captain Piccard *'make it so.'*

Index

Air core frame loops, 179–81
Angle of radiation, 16, 41
Antenna patterns, 38–41
Antenna tuner/tuning unit (ATU), 75–6, 81, 114, 205, 208
Array antennas, 153
Artificial aerials *see* Dummy loads
Artificial ground tuners, 247
Aussie doublet antennas, 109–10
Automatic load control (ALC), 204

BALUN transformers, 65, 67, 78, 91–2, 153, 213–18
  erection, 47
  use on receiving antennas, 108
Beam antennas, 133–4
  impedance matching, 201–202
  VHF/UHF quad, 152
Beamwidth, antenna, 41
Beverage antennas, 82
Bi-square large loop antennas, 151
Bifilar winding, transformer, 216–17

Bobtail curtain array, 168–69
  Thorne, 169–70
Box loops *see* Air core frame loops
Broad-band transmission line transformers, 213
Broadside arrays, 153, 160–1
Bruce array, 165–6

Characteristic impedance, 54
Coaxial cable, 57–8
Coaxial switches, 233, 235
Collinear array, 153, 159
  stacked, 174
  two-element, 157–60
Connections:
  balanced input schemes, 59
  transmission line to receiver/transmitter, 58–60
  wire joints, 61
  wires to antennas/supports, 62–5
Construction hints, 41
Copper-clad steel wires, 51–2
Critical depth, 52

249

Curtain arrays, 164–6, 168–70

D-layer (ionosphere), 8
Decibel (unit), 25
Dellinger fade, 13–14
Delta loop antennas, 135–6
  double, 143–4
  multiband switchable, 139–43
Depth of penetration, 52
Depth winding, wire loops, 179
Dip meters, 225
Dipole antennas, 87
  Aussie doublet, 109-10
  bow-tie, 101–102
  capacitor-tuned wide-band, 110
  delta-fed, 100–101
  folded, 95–7
  half-wavelength, 88–91, 100–101
  inductor-loaded shortened, 111–12
  inverted vee, 98
  multiband, 112–13
  multiband half-wavelength, 92–4
  multiband tuned, 113
  radiation patterns, 87–8
  resonant, 113
  sloping, 99
  vertical, 99–100
  wide-band, 108–9
  wide band half-wavelength, 94, 101
  wider bandwidth folded, 102
  zig-zag, 122–3
Directivity, antenna, 36–7
Diurnal variation, ionosphere, 10
Double-delta loop antenna, 143–4
Doublet antennas, 87
Dummy loads, 232–3

E-layer (ionosphere), 8
Earth *see* Ground
Earth's atmosphere, 2–3
Effective height, loop arrays, 180

Effective radiated power (ERP), 36
Eight-element broadside array, 161
Electrical (E) fields, 24–5
Electrical safety, 42
Electromagnetic radiation, 11–12
  waves, 24–5
End-fire arrays, 153
Erection methods, 44–7, 44–8
EWE antennas:
  basic, 81–2
  bidirectional, 85
  Koontz, 82–4
  reversible, 85

F-layer (ionosphere), 9
Feedpoint impedance, measurement, 222, 226–28
Feedpoint resistance, 226
Flat-top beam *see* W8JK array
Folded dipole antennas, 95–7
  X-array, 171
Four-element broadside array, 160–1
Franklin array, 159

G5RV antennas, 106
  physical structure, 107
Gain, antenna, 36–7
Gamma match, 201, 202
Great circle paths, 15
Grid dip meters, 225
Ground:
  alternative symbols, 19–20
  importance of good, 237–8
  radial system, 245
  traditional solutions, 238–41
Ground rods, multiple, 244
Ground waves, 4
  propagation, 5–6
Ground wire tuner, 246–7

Half-lambda tee antennas, 70–2

Hertzian antennas, 27, 87
  shortened radiator, 124–7
HF band lengths, 158

Impedance bridges, 226–28
Impedance matching, 54, 203–205
  beam antennas, 201–202
  LC networks, 205–8
Impedance matching stubs, 153, 210–13
  see also Q-section
Inductor-loaded antennas, 119
Insulators, center, 64–5
Inverse square law, 23
IONCAP programs, 1
Ionization, atmosphere, 7
Ionosphere, 2, 7
  11 year cycle, 9, 11
  27 day cycle, 9, 10
  disturbances, 12–14
  diurnal cycle, 9, 10
  propagation, 4–5, 7–9
  seasonal cycle, 9, 10
  storms, 14
  use for radio communication, 15–16
  variations, 9–12
Isothermal region, 2
Isotropic radiators, 37–8

K-factor, wave, 17–18

Large loop antennas, 129, 176
Large-scale reflector array, 175
Law of reciprocity, 27
Lazy-H array, 161–4
LC matching networks, 205–08
Lightning arrestors, 66–7
Limited space antennas, 119–22
  radial layout, 128
Line of sight, 6
Linearly loaded tee antennas, 123
Long-wire antennas, 87

Loop antennas, 129
  bi-square, 151
  delta, 135–6
  diamond, 148
  half-delta, 148–9
  half-wavelength, 144–5
  inductor-loaded, 146–8
  preamplifiers, 193
  quad, 129–34
  sharpening, 193
  shielded, 186–91
  testing, 191
  tuning schemes, 181–4
  two-band compact, 150
  using, 191
Low-Noise RX antennas, 83, 84

Magnetic (H) fields, 24–5
Marconi antennas, 27, 69
  basic tee, 70
  folded tee, 80–1
  quarter wavelength vertical, 78–79
  random length, 75–6
  right-angle, 124
  twin-lead, 73–4
Matching stubs *see* Impedance matching stubs
Mechanical integrity, installation, 43–4
MFJ-249 analyzer, 232
MFJ-259 analyzer, 230–42
MFJ-949E tuner, 219
MFJ-956 tuner, 219
MFJ-986 tuner, 219
Multiband tuned doublet antennas, 103, 105
Multipath phenomena, 6

Noise bridges, 228–29

Parasitic elements, 133
Peak reading meter, 224
Phase reversal stubs, 159

Phased vertical arrays, 171–2
Planar winding, wire loops, 179
Polarization, signal, 24–5, 26
Positive ions, 2
Propagation paths, 4–5
 anomalies, 1–2, 3

Q-multiplier, 193–4
Q-section (matching stub), 133, 135–6, 208–10
Quad loop antennas, 129–33
 beam version, 133–4
 fixed, 131
 rotatable, 131, 132
Quarter-wavelength matching sections *see* Q-section
Quarter-wavelength vertical antennas, 78–9

Radial layout, for limited space, 128
Radio signal, 24
Radio waves, 20–3
 frequencey, 22–3
 wavelength, 22–3
Receiver-antenna interactions, 28–9
Reciprocity law, 27
Resistance bridges, 226–8
Resonant frequency (RF):
 measurement, 221
 power meters, 222–4
Roof-mounted antennas, 120–2
Room loop antennas, 136–9

Safety precautions, 68
 altering ground resistance, 242–4
 electrical 42, 68
 ground systems, 247–8
Seasonal cycle, 10
Shielded loop antennas, 186–91
SI units, viii–ix
Signal-to-noise ratio, 25
Six-shooter array, 166–8

Skin effect, 52
Skip communications, 7
Skip zone, 6
Sky wave, 7
Small loop antennas, 176–8
Soil conductivity, altering, 238–44
Solar disturbances, 10–12
 charged particles, 11–12
Solar energy, 11–12
Solar flux index (SFI), 12
Solar noise, 12
Solid wires, 49
Space waves, 4, 5–6
Split frequencies, 178
Spoiler loop, 192
Sporadic-E layer, propagation, 8, 13
Sports fan's loop antenna, 186
Square hobby board loop antennas, 184–6
Stacked collinear arrays, 174
Standing wave ration (SWR), 29–33, 203
 measuring, 33, 35–6
Standing waves, 203
Step attenuators, 235–6
Sterba curtain array, 164–5
Stranded wires, 49–50
Stratosphere, 2
Subrefraction, 18
Sudden ionospheric disturbances (SIDs), 13–14
Sunspots, 12
Super-refraction, 17–18
Surface waves, 4, 5–6
Surge impedance, 54
Swallow tail antennas, 73–5
Symbols, 19–20

Thorne array bobtail curtain (TABC), 169–70
Tilted, center-fed terminated, folded dipole (TCFTFD), 114–17

Transformer loop antennas, 180–1
Transmission lines, 54
  coaxial cable, 57–8
  parallel open-wire, 54, 56
  twin-lead, 56–7
Trap dipole antenna, 112–14
Trifilar winding, transformer, 216–17
Troposphere, 2
  propagation, 4
Tuners:
  commercial, 219–20
  see also Antenna tuner/tuning unit (ATU)
Tuning schemes, loop antennas, 181–4
Twin-lead lines, 56–7
Two-element collinear array, 157–60

Varacators, 183
Velocity factor (VF), conductor, 53–4, 58
Vertical antennas, 119
VHF/UHF band lengths, 159
VOACAP program, 1
Voltage standing wave ratio (VSWR), 32–3, 34–5, 203, 204–5
  analyzers, 230–2

(Voltage standing wave ratio (VSWR) *continued*)
  measurement, 221, 222–4
  performance curves, 221, 222

W8JK array, 172–4
Waves, *see* Radio waves
Windom antennas, 76–8
Wire array antennas, 153–6
Wire joints, 61
Wires:
  basic types, 49–50
  connections to antennas/supports, 62–5
  copper-clad, 51–2
  size to length problems, 52–4
  sizes, 50
Wolf number, 12

Yagi beam antennas, 134, 195–6
  four-element array, 199–200
  six-element array, 200–201
  three-element array, 198
  two-element array, 196–7

Zepp antennas, 81
  double extended, 103, 104, 154
Zig-zag dipoles, 122–3
Zurich smoothed sunspot number, 12